Microbial Surface Components and Toxins in Relation to Pathogenesis

FEDERATION OF EUROPEAN MICROBIOLOGICAL SOCIETIES SYMPOSIUM SERIES

Recent FEMS Symposium volumes published by Plenum Press

1990 • MOLECULAR BIOLOGY OF MEMBRANE-BOUND COMPLEXES IN
PHOTOTROPHIC BACTERIA
Edited by Gerhart Drews and Edwin A. Dawes
(FEMS Symposium No. 53)

1990 • MICROBIOLOGY AND BIOCHEMISTRY OF STRICT ANAEROBES
INVOLVED IN INTERSPECIES HYDROGEN TRANSFER
Edited by Jean-Pierre Bélaich, Mireille Bruschi, and Jean-Louis Garcia
(FEMS Symposium No. 54)

1990 • DENITRIFICATION IN SOIL AND SEDIMENT
Edited by Niels Peter Revsbech and Jan Sørensen
(FEMS Symposium No. 56)

1991 • *CANDIDA* AND CANDIDAMYCOSIS
Edited by Emel Tümbay, Heinz P. R. Seeliger, and Özdem Anǧ
(FEMS Symposium No. 50)

1991 • MICROBIAL SURFACE COMPONENTS AND TOXINS
IN RELATION TO PATHOGENESIS
Edited by Eliora Z. Ron and Shlomo Rottem
(FEMS Symposium No. 51)

1991 • GENETICS AND PRODUCT FORMATION IN *STREPTOMYCES*
Edited by Simon Baumberg, Hans Krügel, and Dieter Noack
(FEMS Symposium No. 55)

1991 • THE BIOLOGY OF *ACINETOBACTER:* Taxonomy, Clinical Importance,
Molecular Biology, Physiology, Industrial Relevance
Edited by K. J. Towner, E. Bergogne-Bérézin, and C. A. Fewson
(FEMS Symposium No. 57)

A Continuation Order Plan is available for this series. A continuation order will bring delivery of each new volume immediately upon publication. Volumes are billed only upon actual shipment. For further information please contact the publisher.

Microbial Surface Components and Toxins in Relation to Pathogenesis

Edited by

Eliora Z. Ron

Tel Aviv University
Tel Aviv, Israel

and

Shlomo Rottem

The Hebrew University–Hadassah Medical School
Jerusalem, Israel

PLENUM PRESS • NEW YORK AND LONDON

Library of Congress Cataloging-in-Publication Data

Microbial surface components and toxins in relation to pathogenesis /
  edited by Eliora Z. Ron and Shlomo Rottem.
       p.   cm. -- (FEMS symposium ; no. 51)
    "Proceedings of a symposium held under the auspices of the
  Federation of European Microbiological Societies, held May 15-19,
  1989, at Ramat Rachel, Israel."--T.p. versc.
    Includes bibliographical references and index.
    ISBN 978-1-4684-8997-2   ISBN 978-1-4684-8995-8 (eBook)
    DOI 10.1007/978-1-4684-8995-8
    1. Bacterial cell surfaces--Congresses. 2. Bacterial toxins-
  -Congresses. 3. Virulence (Microbiology)--Congresses.
  4. Pathogenic microorganisms--Congresses.   I. Ron, Eliora Z.
  II. Rottem, Shlomo. III. Federation of European Microbiological
  Societies.   IV. Series.
    [DNLM: 1. Bacteria--pathogenicity. 2. Bacterial Adhesion-
  -congresses. 3. Toxins--congresses. 4. Virulence--congresses.   W3
  F21 no. 51 / QW 730 M6263 1989]
  QR77.35.M55 1991
  616'.014--dc20
  DNLM/DLC
  for Library of Congress   91-3654
                                                                      CIP

Proceedings of a Symposium held under the auspices of the
Federation of European Microbiological Societies, held May 15–19,
1989, at Ramat Rachel, Israel

ISBN 978-1-4684-8997-2

© 1991 Plenum Press, New York
Softcover reprint of the hardcover 1st edition 1991

A Division of Plenum Publishing Corporation
233 Spring Street, New York, N.Y. 10013

PREFACE

The meeting on "Microbial Surface Components and Toxins in Relation to Pathogenesis" was held on May 15-19, 1989, in the Mitzpe Rachel guesthouse of Kibbutz Ramat Rachel in Israel. Four major topics formed the basis for the meeting: adhesion and colonization; cell invasion and intracellular multiplication; evasion of host defenses; toxins and systemic effects. The presentations clearly show that our understanding of the pathology, pathogenesis and bacteria-host cell inter-action has greatly advanced over the last few years. The contributions to our knowledge on the biogenesis of adhesins and their molecular organization, as well as on the mechanism of adherence to infected target tissue by pathogenic bacteria, have been particularly impressive. Significant progress has been made in defining the nature of pathogenic and cytotoxic factors produced by bacteria, and much has been learned about the biochemical and antigenic modifications occurring in diverse types of host cells upon infection. The discussions of polysaccharide capsules, bacterial endotoxins and secreted toxins illustrated the challenge and the possibilities for vaccine development.

We are indebted to Dr. P. Helen Makela for her continuing encouragement in the initiation and development of the concepts that led to this meeting. Special thanks to the participants and the session chairpersons whose efforts were largely responsible for the vigorous discussion and enthusiasm demonstrated at the various sessions. The meeting was sponsored and supported by the Federation of European Microbiological Societies (FEMS). We are also grateful to the Israeli Microbiological Society, the Hebrew University of Jerusalem and Tel-Aviv University for additional funding. Special thanks to the staff of Mitzpe Rachel for superb organizational efforts, and to Mr. R. Alstock of the American Medical Laboratories (Israel) Inc. and Mr. I. Tsarfaty for their help in preparing the manuscripts.

We hope that this volume will provide a current, exhaustive treatment of the subject for microbiologists, biochemists, immunologists and clinicians.

Eliora Z. Ron
Shlomo Rottem

CONTENTS

I. ADHESION AND COLONIZATION

CHARACTERIZATION AND SURFACE ORGANIZATION

OF *E. COLI* ADHESINS

Klaus Jann and Heinz Hoschützky

Max-Planck-Institut für Immunbiologie
D-7800 Freiburg, FRG

INTRODUCTION

An important initial step in bacterial infections is the adhesion of the pathogenic bacteria to host cells or tissue. This phenomenon has been studied with intestinally pathogenic as well as with extraintestinal and invasive bacteria. Adhesion is mediated by bacterial recognition proteins, which are termed as adhesins. Since adhesive bacteria also induce agglutination of host red blood cells (RBC), the recognition proteins are also called hemagglutinins (1). They are associated with extracellular structures which may have different appearances in the electron microscope. Some of these extracellular appendages are relatively thick and rigid and others are much thinner, flexible and curly. Both can be demonstrated directly by negative staining procedures. A third group of extracellular adhesive structures can only be visualized after stabilization with specific antibodies and has then a capsule-like appearance. These structures, which are schematized in Fig. 1, are termed as fimbriae (rigid, 5-7 nm diameter), fimbrillae (flexible, 2-3 nm diameter) and nonfimbrial (no fine structure demonstrable). Pending their morphological analysis, the term "nonfimbrial" for the latter structures is only tentative. All extracellular adhesive structures are not expressed at growth temperatures of 20°C and below. Accordingly, bacteria grown at these low temperatures are not adhesive and do not agglutinate RBC. With only few exceptions, the adhesins recognize and interact with complex carbohydrates on the surface of host cells (1-5). Some specificities, together with the nature of the respective adhesive structure are shown in Table 1. Minimal receptor requirements are often fulfilled with oligosaccharides of rather simple structures. A number of carbohydrates are yet ill defined. The receptor oligosaccharides are present on glycolipids and/or glycoproteins. Carbohydrate specificity of the recognition proteins are independent of their morphological appearance.

FIMBRIAE ASSOCIATED ADHESINS (FAC)

The adhesive structures canbe easily removed from the

Microbial Surface Components and Toxins in Relation to Pathogenesis
Edited by E.Z. Ron and S. Rottem, Plenum Press, New York, 1991

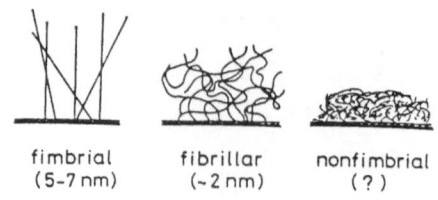

fimbrial   fibrillar   nonfimbrial
(5-7 nm)   (~2 nm)   (?)

adhesin

Fig. 1. Schematic representation of the morphological appearance of *E. coli* adhesins.

bacteria by shearing. They consist of peptide subunits with molecular weights generally in the range of 15-30 kD (6). For some time little progress was made toward the nature of the actual adhesin molecules, and their relation to the observed extracellular structures remained obscure. Research on adhesins gained however momentum from results of genetic analysis reported from several laboratories (7-11). It was found with *E. coli* expressing P, S, and type 1 fimbriae that fimbriation and adhesiveness are determined in a large gene complex which could be cloned. Both properties can mutate independently and this can result in afimbrial but adhesive mutants or in fimbriated but non adhesive mutants. From these findings one can conclude that fimbriae and adhesins are distinct entities which one should be able to separate.

We have adopted this working hypothesis, and in an extensive investigation we were able to isolate the Gal-Gal (P) and the Sia- Gal (S) specific adhesins from the respective fimbriae--adhesin complexes (FAC)(12,13). Purified FAC were dissociated with Zwittergent at elevated temperature. The fimbriae from

Table 1. Receptor requirements of some adhesins from *Escherichia coli*.

| RECEPTOR | | |
DOMAINE	ON	ADHESIVE ORGANELLE
$(\alpha\text{-Man})_n$	GLYCOPROTEINS	FIMBRIAL (TYPE 1), NONFIMBRIAL
$\alpha\text{-Gal-1,4-}\beta\text{-Gal}$	GLYCOLIPIDS	FIMBRIAL (P)
$\alpha\text{-GalNAc-1,3-}\alpha\text{-Gal}$	GLYCOLIPIDS	FIMBRIAL (PRS)
$\alpha\text{-Sia-2,3/6-}\beta\text{-Gal}$	GLYCOPROTEINS	FIMBRIAL (S)
$\beta\text{-GalNAc-1,4-}\beta\text{-Gal-1,4-}\beta\text{-Glc}$ $\quad\quad\quad\quad\overset{3}{\underset{2}{\mid}}$ $\quad\quad\quad\quad\text{Sia}$	GLYCOLIPIDS	FIMBRIAL (CFA/I)
Gal, GalNAc	GLYCOLIPIDS	FIBRILLAR (K88, K99)
Gal, GalNAC, SERINE	GLYCOPROTEINS (M/N)	NONFIMBRIAL

Table 2. Properties of S- and P- specific adhesins and their
         fibrillins.

PROPERTY	S		P (F7$_1$)	
	FIM (A)	ADH (S)	FIM (A)	ADH (G)
APPAR. M.W.	10^4 K	10^4 K	10^4 K	10^4 K
SUBUNIT W.	16.5 K	14.5 K	22 K	35 K
I.E.P	6.0	4.7	4.8-5	4.8
CYSTEINS	2	2	2	8
METHIONINS	1	2	2	4
AGGL./ADH.	-	+	-	+
INHIBITOR	n.d.	Sia-Gal-Glc	n.d.	Gal-Gal-Glc

which the adhesins had been removed could, like FAC, be
precipitated with lithium chloride and the adhesins were
purified from the supernatants. Some properties of the
adhesins and the respective subunits are shown in Table 2.
The adhesin from S fimbriae is called the S and that from P
fimbriae is called the G protein. The mass ratio of fimbriae
to their adhesin is several hundred. It should be mentioned
that the native fimbriae adhesin complexes contain, besides
the adhesins, additional minor components. In the case of
P-FAC these are called the E and F proteins (9,14) and in the
case of S-FAC they are called the G and H proteins (12a).
These minor fimbrial proteins are needed for the assembly of
complete functional FAC. The adhesins differ from the fimbrial subunits in molecular-weight, amino acid composition and
sequence, isoelectric points (surface charge) and, importantly, in their capacity to adhere to host cells. To test the
specificity of adherence, we used an assay in which the
immobilized adhesins were incubated with RBC. The extent of
adhesion was measured by lysis of the adherent RBC and by
determination of the released haemoglobin (13). The hem-
adhesion was specifically inhibited with α-Gal-(1,4)-
Gal-(1,4)-Glc.

A differentiation between fimbriae and adhesins was also
possible with the use of monoclonal antibodies. One of the
adhesin specific antibody inhibited hemagglutination and all
the fimbrillin specific antibodies did not inhibit (12). The
antibodies could be used in immunoelectron microscopy to show
that the adhesins are located at the tip of fimbriae.

We found that purified P-FAC react in ELISA with a
monoclonal anti-LPS antibody. Furthermore, P-FAC are cytotoxic for stimulated phagocytes, as tested in chemiluminescence.
FAC from mutants lacking the P adhesin (G protein) or fimbriae after removal of the P adhesin are not cytotoxic. Our
monoclonal anti- LPS antibody was used in the laboratory of
W. Hoesktra to label the tips of P-FAC with the immunogold
technique. FAC from G- mutants could not be labelled. Linder
et al. (15) found that P- FAC induce an inflammatory response
in mice. Similar findings were not obtained with S-FAC. From

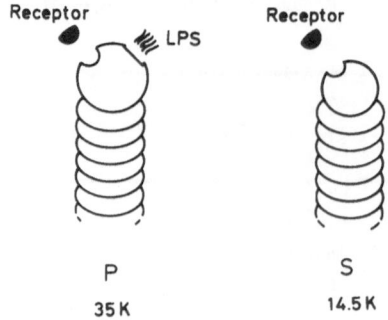

Fig. 2.   Schematic representation of receptor binding site
and LPS-binding site of the 35K P-adhesin and of
receptor binding site only on the 14.5K S-adhesin.

this we assume as a working hypothesis that LPS can specifi
cally bind to the P adhesin.

Since the P adhesin is significantly larger (35 kD) than
the S adhesin (14.5 kD), it is likely that the S adhesin may
lack a LPS binding site which is present in the P adhesin
(Fig. 2). If this assumption were true, the P adhesin would
be a lectin with the properties of a LPS binding protein. In
many laboratories, including our own, the hemagglutinating
capacity of isolated adhesive fimbriae was found to be errat-
ic and not always easy to reproduce. We observed that the pH
value of the solute is important for agglutination. At pH 7,
purified P and S fimbriae adhere to cells but do not hemagglutinate, whereas at pH 5-5.5 adherence and hemagglutination
occurs. The suspensions are more turbid at pH 5 than at pH 7.
We interpret this, as shown schematically in Fig. 3, such
that at lower pH the fimbriae from hemagglutinating aggregates. The pH dependence of hemagglutination can thus be
taken as an indication of monovalency of P and S fimbriae.
Similar results as described for the P and S fimbriae were in
principle also reported with the mannose specific type 1
fimbriae.

NONFINMBRIAL ADHESINS (NFA)

Unfimbriated adhesive E. coli have been described earlier and studies on nonfimbrial adhesins are being carried out
in several laboratories (16-23). From some strains of adhesive uropathogenic E. coli, which did not exhibit fimbriae,
we have isolated several nonfimbrial adhesins (21-23). They
were extracted from the bacteria at elevated temperature and
purified by fractional precipitations and centrifugations,
followed by anion exchange HPLC. The NFAs differed in surface
charge (HPLC elution profile, isoelectric point) and molecular weight of their subunits (SDS- PAGE). The apparent molecular weights of all NFA were above 10^7 D, indicating that
they consisted of more than 100 apparently identical subunits. Amino acid sequencing (F. Lottspeich, MPI Martinsried,
FRG) showed as only positive compositional correlation with
other proteins a 70% homology within the 20 amino terminal

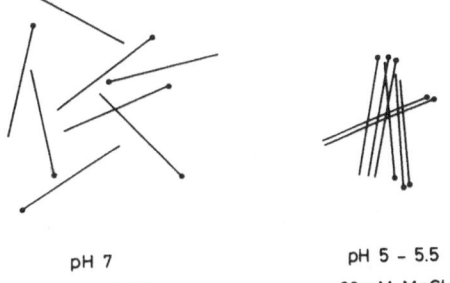

pH 7

‹ 1 mM MgCl₂

pH 5 – 5.5

~ 20 mM MgCl₂

Fig. 3.   Schematic representation of the influence   of pH on
hemagglutination capacity of fimbriae   via aggrega-
tion.

amino acids   of the fimbrillar   K88 antigen of   calf   enteropathogenic *E. coli*.

    In contrast to the P and   S fimbriae, all NFA adhere   to
and agglutinate only   human cells. Interestingly, hemagglutination is equally   strong at pH 7   and pH 5, which   indicates
polyvalency of the NFA. The carbohydrate   receptor of all NFA
is contained in glycophorin A. Some adhesins prefer Aᴹᴹ , one
(NFA3)   recognizes Aᴺᴺ, and others do not seem to differentiate.

    We have prepared monoclonal antibodies against NFA, most
of which are antiadhesive (23). ELISA studies showed that the
NFA are   distinct antigens,   although some   serological cross
reactivity can be observed.

    To   elucidate   the   superstructure of   the   nonfimbrial
adhesins, we have begun   their analysis with light scattering
measurements.   First results   indicate that   also NFA   have a
fibrillar   structure, comparable in   linear mass   density and
strand   flexibility to   the K88   fibrillae. The   capsule like
structure observed in immunoelectron microscopy (Fig. 4), may
be due to interstrand   fixation of fibrillae with the   formation of microgel-like structures.

MORPHOLOGY OF *E. COLI* ADHESINS

    On the   basis of electron   microscopy, light scattering,
pH dependence of hemagglutination and disintegration studies,
it appears that all   *E. coli* adhesins known today   are either
associated with or form   fimbrial of fibrillar structures. In
Fig. 5 the adhesins of   P, S, CFA/I, K88, α-3000 from   a uropathogenic   strain   and NFA   are compared. Some of   them are
fimbrial (P, S, CFA/I, 0-3000) and others are   fibrillar (K88
and NFA).   Of these, only the P, S, and also type 1, fimbriae
have distinct   adhesins (recognition   proteins) at   their tip
and the fimbriae   themselves are only   supporting structures.
These   FAC   are monovalent.   The   other   filamentous adhesive
complexes consist   entirely of adhesin subunits.   They may be
arranged so that only   the tip located adhesin is   accessible
for receptor   recognition (fimbrial CFA/I, fibrillar   K88) or
so that all ormany subunits are accessible (fimbrial   CFA/I,

7

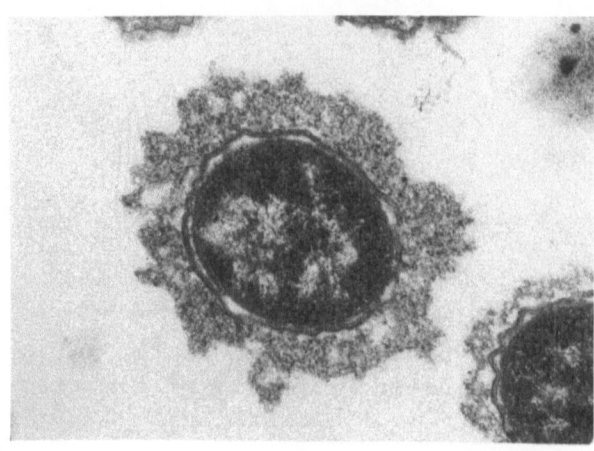

Fig. 4. Electronmicrograph of a thin section from Epon--embedded *E. coli* 083:H⁻ expressing the nonfimbrial adhesin NFA-1. (J. Golecki, University of Freiburg).

fibrillar NFA). The latter ones are multivalent. Very recently we have studied the thermal stability of *E. coli* adhesins. Whereas the type 1, P and S fimbriae are stable at 100°C, the others depolymerize to the monomers and multiples of it. Without additional disintegration with SDS the heat depolymerized samples exhibit ladder-like patterns in SDS-PAGE. Practically the same results are obtained with the NFA. The CFA/I fimbriae and K88 fibrillae seem to be even more thermosensitive. The oligomeric subunits are stable for some time and can grow back into larger structures under defined conditions. These results indicate difference in the strength of the non-covalent bonds between the fimbrial and fibrillar subunits. They open the way to a more comprehensive study of bacterial adhesins, their molecular organization and possibly also their functions *in vivo*.

CO-EXPRESSION OF ADHESINS AND POLYSACCHARIDE CAPSULES

Adhesins, mediating bacterial attachment and capsules, mediating resistance to host defense, are important virulence determinants. It therefore stands to reason that both structures are expressed on the same cells. However, as intestinal *E. coli* do generally not have capsules, a co-expression of adhesins and capsules can only be expected with extra-intestinal *E. coli*, and in principle there is co-expression of P, S, of the fibrillar NFA with polysaccharide capsules. Since adhesins may switch from an "on" state to an "off" state and mutation from capsule plus to capsule minus is not infrequent we wanted to know whether both these structures are occurring on one cell. Immunoelectron microscopy indicated that cellular co-expression of fimbriae and polysaccharide capsules is not frequent. In contrast, the nonfimbrial adhesins and polysaccharide capsules are co-expressed more frequently. These extra-cellular compartments appear as two types of capsules after stabilization with monoclonal antibodies. They could also be differentiated with the antibodies

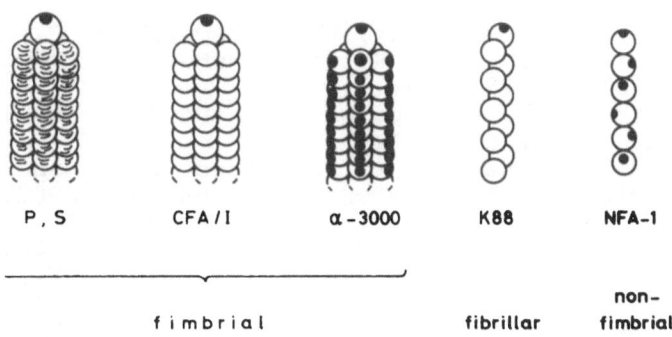

P, S   CFA/I   α-3000   K88   NFA-1

|_____|
          fimbrial   fibrillar   non-
                                                fimbrial

FORMS OF ADHESINS

Fig. 5.   Schematic representation of the molecular organiza-
          tion of *E. coli* adhesins.

with the immunogold technique. The adhesin was the outer-most
layer surrounding the polysaccharide capsule and connected
with the cell surface through thread like structures (24). A
possible participation of both extracellular structures in
adhesion or opsonophagocytosis has not yet been studied.

ACKNOWLEDGEMENTS

      This work was supported by the Bundesministerium fur
Forschung und Technologie (BMFT-DFVLR) and by the Deutsche
Forschungsgemeinschaft (DFG).

REFERENCES

1.   Duguid, J.P. and Old, D.C. 1980. Bacterial Adherence,
      E.H. Beachey ed., Capman & Hill, London, 185-217.
2.   Leffler, H. and Scanborg-Eden, C. 1980. FEMS Microbiol.
      Lett. 8:128-134.
3.   Korhonen, T.K., Vaisinen-Rhen, V., Rhen, M., Pere, A.,
      Parkkinen, J., and Finne, J.J. Bacteriol. 159:762-766.
4.   Vaisanin, V., V., Korhonen, T.K., Jokinen, M., Gahmberg,
      C.G., and Ehmoholm, C. 1982. Lancet, i,1192
5.   Ofek, I. and Sharon, N. 1988. Curr. Top. Microbiol.
      Immunol. 151:91-114.
6.   Jann, K., Jann B., and Schmidt, G. 1981. FEMS Microbiol.
      Lett. 11:21-25.
7.   Minion, F.C., Abraham, S.N., Beachey, e.H., and Goguen,
      J.D. 1986. J. Bacteriol. 165:1033-1036.
8.   Lund, B., Lindberg, F.P., Baga, M., and Normak, S. 1985.
      J. Bacteriol. 162:1293-1301.
9.   Lindberg, F.P., Lund, B., and Normak, S. 1986. Proc.
      Natl. Acad. Sci. USA 83:1891-1985.
10.   vanDie I., Zuidweg, E., Hoekstra, W., and Bergmans, H.
      1986. Microbiol. Pathog. 1:51-56.
11.   Hacker, J., Schmidt, G., Hughes, C., Knapp, S., Marget,
      M., and Goebel, W. 1985. Infect. Immun. 47:434-440.
12.   Moch, T., Hoschutzky, H., Hacker, J., Kroncke, K.-D.,
      and Jann, K. 1987. Proc. Natl. Acad. Sci. USA
      84:4362-3466.

12a. Schmoll, T., Hoschutzky, H., Morschhause, J., Lottspeich, F., Jann, K., and Hacker, J. 1989. Mol. Microbiol. 3:1735-1744.

13. Hoschutzky, H., Lottspeich, F., and Jann, K. 1989. Infect. Immun. 57:76-81.

14. Lund, B., Lindberg, F.P., Marklund, B.I., and Normak, S. 1987. Proc. Natl. Acad. Sci. USA 84:5898-5902.

15. Linder, H., Engberg, I., Matzby Baltzer, I., Jann, K., and Svanborg-Eden, C. 1988. Infect. Immun. 56:1309-1313.

16. Orskov, I., Birch-Andersen, A., Duguid, J.P., Stenderup, J., and Orskov, F. 1985. Infect. Immun. 47:191-200.

17. Forestier, C., Welinder, K.G., Darfeuille-Michaud, A., and Klemm, P. 1987. FEMS Microbiol. Lett. 40:47-50.

18. Williams, P.H., Knutton, S., Brown, M.G.M., Candy, D.A.C., and McNeish, A.S. 1984. Infect. Immun. 44:592-598.

19. Knutton, S., Lloyd, D.R., and McNeish, A.S. 1987. Infect. Immun. 55:86-92.

20. Walz, W., Schmidt, A., Labigne-roussel, A.F., Falkow, S., and Schoolnik, G. 1985. Eur. J.Biochem. 152:315-321.

21. Goldhar, J., Parry, R., Golecki, J.R., Hoschutzky, H., Jann, B., and Jann, K. 1987. Infect. Immun. 55:1837-1842.

22. Grunberg, J., Perry, R., Hoschutzdy, H., Jann, B., Jann, K., and Goldhar, J. 1988. FEMS Michrobiol. Lett. 56:241-246.

23. Hoschutzdy, H., Nimmich, W., Lottspeich, F., and Jann, K. 1989. Microbiol. Pathog. 6:351-359.

24. Kroncke, K.D., Orskov, I., Orskov, F., Jann, B. and Jann, K. Infect. Immun. 58:2710-2714.

BINDING OF ENTEROBACTERIAL FIMBRIA TO PROTEINS OF BASEMENT MEMBRANES AND CONNECTIVE TISSUE – A NOVEL FUNCTION FOR FIMBRIAE

Timo K. Korhonen, Benita Westerlund, Ann-Mari Tarkkanen, Timo Sareneva, Ritva Virkola, Pentti Kuusela, Harry Holthofer, Irma van Die, Wiel Hoekstra, Bradley L. Allen, and Steven Clegg

Departments of General Microbiology, and Bacteriology and Immunology, University of Helsinki, Helsinki Finland; Department of Molecular Cell Biology University of Utrecht, Utrecht, The Netherlands; Department of Microbiology, University of Iowa, Iowa USA

INTRODUCTION

Fimbriae are important virulence factors in a number of enterobacterial infections. It is generally thought that they increase the pathogenic potential of bacteria by mediating adherence to glycoconjugates of host epithelia, thus giving resistance to mechanical clearance defense systems of the body. Indeed, the various fimbrial types found on *Escherichia coli* causing extraintestinal infections in humans (reviewed in 1-3) have been shown to exhibit a distinct tissue tropism in their adhesion to mammalian epithelia (2-9). Some of these fimbriae also interact selectively with soluble compounds found in normal human urine (10), and it seems that the pathogenicity of fimbriae in urinary tract infections depends both on the availability of fimbrial receptors or binding molecules in tissues and on the absence of soluble adherence inhibitors in normal human urine (reviewed in 2,3).

The virulence functions of fimbrial filaments are usually associated with their binding to epithelial cells via lectin-like moieties, regardless of their presence as major (11-13) or minor (14-16) components of the filaments. Recently, a number of enterobacterial isolates have been found, which interact with solubilized proteins of basement membranes (BM) or interstitial connective tissue (CT) (17,18), raising the question on the role of fimbriae in these interactions (19). By using histological methods to localize fimbrial binding sites in frozen tissue to BM or CT in human kidney and urinary bladder (20,21). These fimbriae recognize different components of BM or CT, and an intriguing property of at least one these interactions (20) is that the receptor-binding property is mediated by a portion of the fila-

Microbial Surface Components and Toxins in Relation to Pathogenesis
Edited by E.Z. Ron and S. Rottem, Plenum Press, New York, 1991

11

ment distinct from a minor protein previously identified as a specific lectin (14,15). Therefore fimbriae appear as multi-functional organelles, with lectin- dependent and lectin--independent functions. In this chapter we will summarize our recent studies on binding of fimbriae to proteins of basement membranes and connective tissue.

BINDING OF FIMBRIAE TO BASEMENT MEMBRANES IN TISSUE SECTIONS

We have systematically studied the tissue-substructure specificity of all major fimbrial types identified on entero-bacterial strains associated with urinary tract infections or neonatal meningitis. The P and S fimbriae seem to bind effec-tively to epithelial and endothelial cells in the kidney as well as in the neonatal rat brain (6,7,9), and it seems likely that such a binding contributes to their pathogenic function in pyelonephritis and neonatal meningitis, respec-tively. However, certain fimbrial types bind specifically to BM or interstitial CT in kidney and other parts of the uri-nary tract. These adhesins include the O75X adhesin (22) (also called the Dr adhesin; 23), the type-3 fimbriae of *Klebsiella* (24) and fimbriae from *Proteus* strains associated with urinary tract infections (25).

Kidney offers a useful target tissue for studies on bacterial adherence to BM and CT. Besides being the relevant tissue for studies on uropathogenic bacteria, it offers a well-studied model for the structure of BM and CT and for the distribution of various glycoconjugates in different tissue subcompartments (26,27).

The O75X adhesin is a fimbria-like organelle associated with uropathogenic *E. coli* strains with the O-group 75 (28). The adhesin has also been termed the Dr adhesin due to its ability to hemagglutinate human erythrocytes carrying the Dr blood group antigen (23). The adhesin has been purified and characterized (28) and the gene cluster encoding the adhesin was recently cloned from the chromosome of an uropathogenic *E. coli* strain (13). A wild type strain expressing the O75X adhesin, bind specifically to BM in frozen sections of kidney (5,29). The binding sites for this adhesin have also been mapped along the entire canine urinary tract (22).

The purified O75X adhesin binds strongly to tubular BM, arterial walls and Bowmann's capsule around glomeruli, but not to tubular epithelial cells or endothelium (Fig. 1A). A weak staining of the glomerular mesangium is also evident (Fig. 1A). In urinary bladder, the adhesin shows strong binding to CT between smooth muscle cells (Fig. 1B).

Type-3 fimbriae are found on nearly all *Klebsiella* isolates from patients with urinary or respiratory tract infections (30). This fimbrial type is found also in other enterobacterial species, although less frequently, and is characterized by conserved structure in members of the *En-terobacteriaceae* (31). In frozen sections of human kidney, the purified type-3 fimbriae bind specifically to intersti-tial CT, tubular BM, arterial walls and Bowman's capsule (Fig. 1C). A recombinant *E. coli* strain expressing the type-3-fimbrial gene cluster of *Klebsiella pneumoniae* (16) specif-

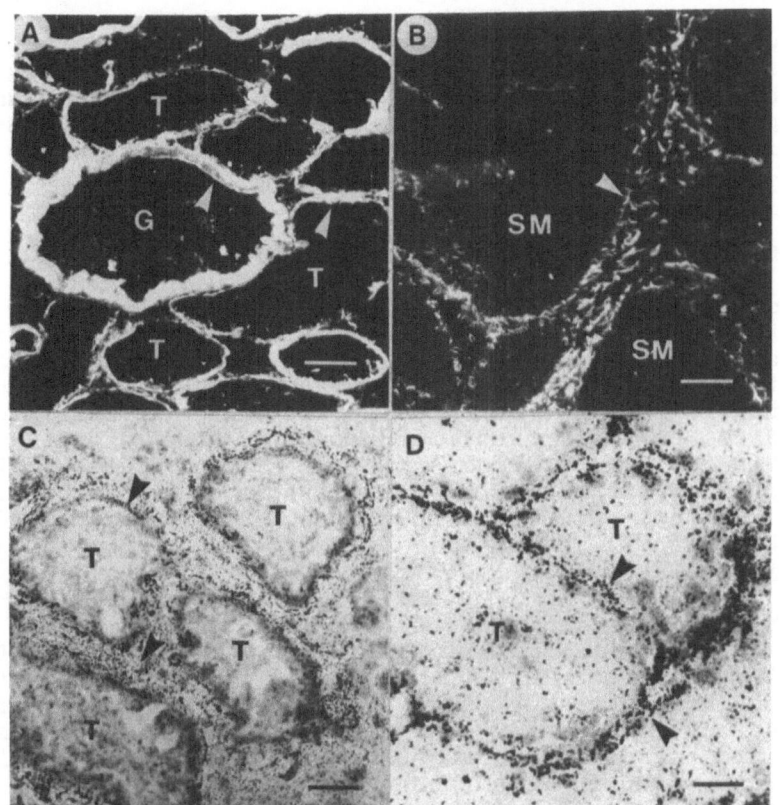

Fig. 1.　　Binding of purified enterobacterial fimbriae to BM
and CT in frozen tissue sections. Binding was visualized by indirect immunofluorescence in (A,B) and
by indirect immunogold silver staining in (C,D).
(A) Binding of O75X adhesin to tubular BM and
Bowmann's capsule; symbols T and G indicate tubuli
and glomerulus. (B) Binding of O75X to CT (arrowheads) between smooth muscle cells (SM) in human
urinary bladder. Binding (arrowheads) of type-3
fimbriae of *K. pneumoniae* (C) and of *P. mirabilis*
fimbriae (D) to BM and CT in human kidney are also
shown. Bars, in A-C 25 μm, in D 10 μm.

ically adheres to the same tissue domains (24), confirming
that the observed binding is due to type-3 fimbriae.

Fimbriae have been associated with uropathogenicity of
Proteus (32), and we have recently detected a wide array of
tissue tropism by fimbriae of *Proteus* strains (25). Some
Proteus fimbriae bind exclusively to epithelial and endothelial cells of human kidney, whereas those from *Proteus mirabilis* strain 2456 bind selectively to BM and interstitial CT
(Fig. 1D). The 2456 cells adhere to the same tissue domains,
and the bacterial adhesion is specifically inhibited by Fab
fragments of anti-fimbriae antibodies (21). Fimbriae serologically similar to those of strain 2456 commonly occur in
uropathogenic *Proteus* strains, which suggests that the ability to bind to BM and CT is a common feature among these
bacteria.

The examples described above indicate that many fimbrial

types of pathogenic enteric bacteria have adapted the capacity to specifically bind to BM and CT in human tissues and that the functions of fimbriae are not limited to adherence upon epithelial surfaces. It is noteworthy that in *Klebsiella* and *Proteus* such a tissue-binding specificity seems to be common in pathogenic strains.

BINDING OF FIMBRIAE TO GLYCOPROTEINS OF BM AND CT

To identify putative fimbrial receptors in BM and CT, we have tested *in vitro* the possible interaction of fimbriae with isolated components of the renal BM or interstitial CT. The methods used include adherence of recombinant and wild-type strains to glass surfaces coated with the test proteins, a modified enzyme-linked immunosorbent assay (ELISA) to measure binding of purified fimbriae to proteins coated on microtiter wells, affinity chromatography of bacterial extracts through columns with immobilized glycoprotein, and quantitation of binding of radiolabelled glycoproteins to bacterial cells (20, 21, 24).

The selective binding of the fimbriae to immobilized BM and CT proteins is illustrated in Fig. 2. *E. coli* HB101 is a commonly used recipient strain in cloning experiments and does not adhere to any of the tested BM or CT proteins. In contrast, HB101 harboring the plasmid pBJN406 adheres strongly to glass slides coated with type IV collagen (Fig. 2). The recombinant strain HB101(pBJN406) expresses the O75X (Dr) adhesin (13). Similar adhesion specificity is seen with a wild-type O75-positive strain (20,21). On the other hand, *E. coli* HB101(pFK12) expressing the type-3 fimbriae of *Klebsiella* (16) adheres selectively to glass surfaces coated with type I and type V collagens (Fig. 2; 24). The tissue-binding tropism of the two fimbriae in frozen kidney sections is very similar (Figs. 1A and 1C), but these bacterial structures specifically recognize different components of BM, i.e. the type IV versus the type V collagens. It seems likely that the ability of HB101(pFK12) to recognize type I collagen (Fig. 2) may explain binding of type-3 fimbriae to interstitial CT in tissue sections (Fig. 1C).

Strain HB101 harbouring the plasmid p110-75 that encodes P fimbriae (15) adheres to immobilized fibronection (FN) (Fig. 2.) The active domains in FN are located in the aminoterminal 30K and the carboxyterminal 120K fragments of the molecule (20). We have recently observed a number of *Salmonella* strains capable of adhering to immobilized laminin (Fig. 2); the bacterial adhesin involved has recently been identified as a specific fimbria.

We have confirmed the role of fimbriae in these interactions by testing with ELISA technique the binding of purified bacterial fimbriae to BM and CT proteins immobilized on microtiter plates (20, 21, 24). Fig. 3 shows binding of the O75X adhesin to microtiter plates coated with various test proteins. The O75X binding is determined as a function of either the coating concentration (Figs. 3A and 3B) or the concentration of O75X used in the test (Figs. 3C and 3D). In both cases, a specific and saturable reaction with type IV collagen is evident. The interaction of O75X and type IV

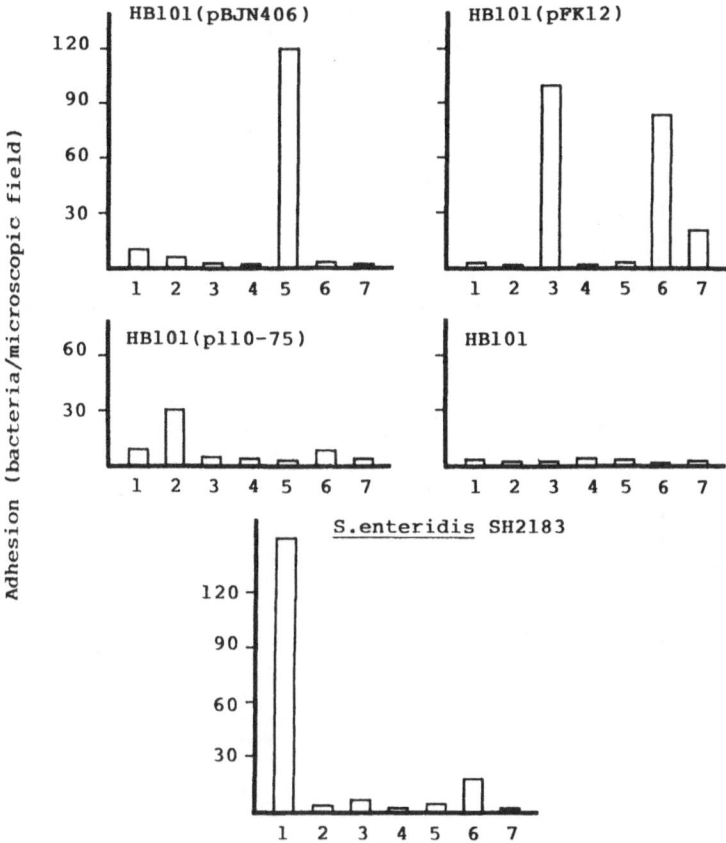

Fig. 2.   Attachment of bacteria to isolated BM and CT proteins adsorbed on glass slides. The coating was
with (1) laminin, (2) fibronectin, (3) type I collagen, (4) type III collagen, (5) type IV collagen,
(6) type V collagen, and (7) bovine serum albumin.
Coating was 1 pmol of the test proteins except for
E. coli strain HB101(pFK12) (5 pmol) and for *Salmonella enteritidis* SH2183 (2.5 pmol). The bacterial
density in the assay varied from 5×10^7 to 3×10^{10} cells per ml in different strains. *E. coli*
HB101(pBJN406) expresses type O75X (Dr) adhesin and
adheres strongly to type IV collagen. HB101(pFK12)
expresses type-3 fimbriae of *Klebsiella* and interacts strongly with type I and type V collagens.
HB101(p110-75) expresses P-fimbriae genes and is
bound to fibronectin. The plasmidless strain HB101
does not interact with any of the immobilized proteins. *S. enteritidis* SH2183 shows a strong interaction with laminin. The calculated microscopic
field was 4.8×10^3 μm².

collagen could also be demonstrated by affinity chromatography on immobilized collagen and by binding of radiolabelled
type IV collagen to bacterial cells expressing the O75X
adhesin (21). The active region in type IV collagen is located in the 7S aminoterminal domain (21). The interaction is

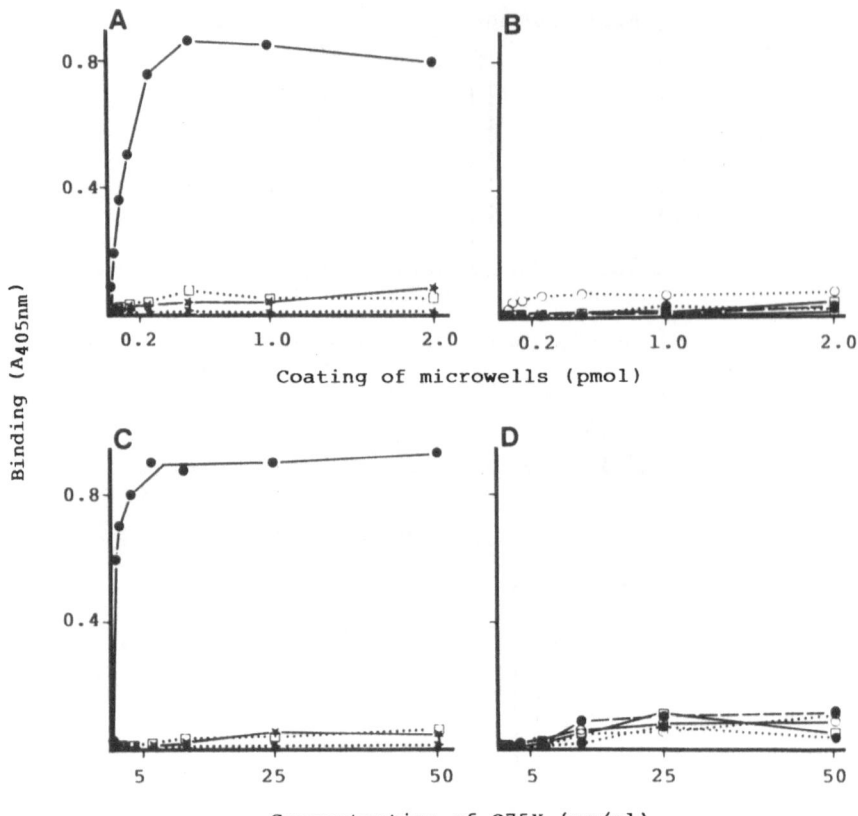

Fig. 3.   Binding of the O75X adhesin to immobilized BM and
          CT proteins. (A,B) Binding of O75X to microtiter
          plates coated with increasing concentrations of the
          test proteins. (C,D) Binding in increasing concen-
          trations of O75X to microwells coated with o=0.5
          pmol of test proteins. Binding was quantitated by
          anti-O75X antibodies and phosphatase-conjugated
          secondary antibodies. Symbols: (o...o), type I
          collagen; (o...o), type III collagen; (o---o),
          procollagen type III; (o---o), type IV collagen;
          (˜---˜), type V collagen; (x---x), type VI colla-
          gen; (˜...˜), laminin; (o- - -o), fibronectin; and
          (x...x), fetuin. Reproduced from ref. 21 with the
          permission of Blackwell Scientific Publications.

inhibited by small amounts of chloramphenicol (21), which
also inhibits binding of O75X to human erythrocytes and is
thought to mimic tyrosine-containing structures in the erythrocyte receptor for O75X (23).

     Type IV collagen is composed of three helically orga-
nized peptide chains. The individual separated chains do not
show interaction with the O75X adhesin, suggesting that the
active site is conformation-dependent and involves two or
three peptide chains of collagen molecule. The finding that
chloramphenicol effectively inhibits the binding further
suggests that the interaction is based on a protein - protein

interaction. Type IV collagen is specific for BM (33) and the binding of O75X to this collagen type is in accordance with the localization of the O75X binding sites in kidneys (Fig. 1A).

In a similar test system, we have detected a specific binding of purified type-3 fimbriae to immobilized type I and type V collagens (24). This finding is in accordance with the results employing strain HB101(pFK12) as shown in Fig. 2. The tissue distribution in kidney of type I and type V collagens (27) is similar to that of type-3-fimbrial binding sites shown in Fig. 1C, suggesting that the observed interactions can take place in situ as well. We have recently shown (24) that this interaction is mediated by the MrkD minor lectin molecule of the type-3-fimbrial filament (16). Noteworthy in this interaction is that it does not take place with soluble collagens. This is most likely due to conformational differences between the soluble and immobilized forms of the collagen molecules, which are known to form highly ordered crystalline arrays in tissues.

A third interaction that we have characterized in some detail is the binding of the P fimbriae of pyelonephritogenic E. coli to FN (20; Fig. 2). Purified P fimbriae retain the capacity to bind FN and its 30K and 120K fragments, but only in their immobilized forms. As in the type-3 fimbriae/collagen interaction, this may be explained by the known conformational differences between the soluble and immobilized forms of FN. The aminoterminal 30K fragment of FN, which is active in the P fimbria/FN interaction (20), is better exposed in the immobilized form of FN. Binding of FN to P fimbriae does not involve the cell-binding RGDS sequence of FN nor the αDGal(1-4)βDGal-binding G protein of the P-fimbrial filament (20). Our ongoing experiments have indicated that this binding involves at least one of the other minor proteins of the P-fimbrial filament, the E protein. The P fimbriae/FN interaction is the first described binding property in a fimbria that is independent of an identified minor lectin protein of the filament. At present we cannot distinguish whether the E protein is another lectin-like protein on the P-fimbrial filament, with a different binding specificity from the G protein, or whether the E protein is a target for fibronectin, known to possess multiple binding specificities.

The BM and CT receptors for the Proteus fimbriae (Fig. 1D) are not known yet. We do not at the present know whether the strong interaction between laminin and Salmonella cells (Fig. 2) is seen in tissue sections as well.

CONCLUSIONS

The examples described in this chapter indicate that many members of the Enterobacteriaceae are able to use fimbriae to interact with protein components of BM and CT. The finding that these interactions are seen in tissue sections as well (Fig. 1), indicates that the BM and CT proteins are true tissue targets for these enterobacterial adhesion proteins. Thus the functions of fimbriae in enterobacterial infections are not restricted to binding to epithelial surfaces, which is thought to be important at early stages of the infections. This idea is supported by the expression of certain fimbrial types in body fluids at late stages of a systemic *E. coli* infection (34).

In type-3 fimbriae, the interaction with collagens involves the minor lectin molecule of the fimbrial filament, whereas in P-fimbrial binding to fibronection the αDGal(1-4)βDGal-binding *fstG* protein is not involved. Until now, it has been thought that the function of the accessory proteins in fimbrial filaments is to express the lectin protein beyond the capsular and lipopolysaccharide layers of the bacterial cell surface, which is needed for active function of the lectin protein in wild-type strains (35). Our results suggest that the other proteins of the filament may also have binding properties of their own, i.e. the fimbrial filaments are multifunctional with lectin-dependent and lectin-independent functions. We have recently detected another lectin-independent property of fimbria in binding of plasminogen (36), which binds to several *E. coli* fimbriae via its lysin-binding domains. Plasminogen is a precursor of plasmin, a basement-membrane-degrading enzyme, and requires immobilization to become activated by proteolysis. *E. coli, Salmonella* and *Yersinia* are known to possess a plasminogen activator (37), and it could be that fimbrial interactions between proteins of BM and plasminogen help the bacteria to invade human tissues.

The interaction of enterobacterial fimbriae with BM and CT proteins seems highly specific. The bacterial proteins interact selectively with one or two collagen types, fibronection or laminin (Fig. 2). Laminin, type IV collagen, and type V collagen are components of BM (27), and it is remarkable that *E. coli, Salmonella* and *Klebsiella* specifically interact with only one of these BM proteins.

It is evident that some of the interactions described in this chapter are highly dependent on the conformation of the BM and CT compounds functioning as target molecules. The type-3 fimbria binds only to the immobilized forms of type I and type V collagens (24), and similarly P fimbriae only interact with the immobilized FN (20). A number of binding properties of FN are dependent on its conformation (see references in 16), and we propose that the failure of soluble FL and collagens to bind to P and type-3 fimbriae results from conformational properties of FN and collagens. In tissues, FN and collagens form highly organized and interwoven structures which may interact with the other components of BM and CT (27,33). it is conceivable that the immobilized forms of BM and CT proteins are more relevant for bacterial adherence to tissues than are the soluble forms, which may even act as inhibitors of bacterial adherence in body fluids.

The biological significance of the interactions between fimbriae and components of BM and CT is not established as yet. The type-3-fimbriate klebsiellas cause urinary and respiratory tract infections mainly in elderly and compromised hosts who may have limited tissue damage. This loss of the normal mucosal defence system may result from many causes including mechanical trauma of intubation or urinary catheterization, viral infection of the upper respiratory tract, aspiration of acidic gastric contents, and inflammatory processes in general, It can be speculated that ability to bind to BM and CT is a useful property for the invading bacteria in such circumstances. Finally, laminin receptors are known to play a role in spreading of metastatic tumor cells through basement membranes (38), and bacteria might use analogous mechanisms for invasion.

SUMMARY

Several enterobacterial species and their purified fimbrial proteins possess the capacity to interact with basement membranes or connective tissue element in frozen sections of mammalian tissues. The currently recognized fimbrial types with this property include the 075X adhesin of uropathogenic *Escherichia coli*, the type-3 fimbria of pathogenic *Klebsiella* and other members of the *Enterobacteriaceae*, and fimbriae of uropathogenic *Proteus* strains. These fimbrial types utilize different components of basement membranes and/or connective tissue as target molecules. The 075X adhesin binds strongly to type IV collagen, and the type-3 fimbriae to type I and type V collagens. Moreover, certain *Salmonella* strains show strong and specific binding to laminin. In the type-3 fimbria, the interaction with collagens involves the MrkD minor lectin protein of the filament, whereas the P fimbria of pyelonpehritogenic *E. coli* interacts with fibronectin by a mechanism that does not involve the lectin protein of the fimbrial filament. Enterobacterial adhesion proteins thus have evolved varying strategies to interact with proteins of basement membranes and connective tissue. These interactions may play a role in bacterial colonization of locally damaged mucosal surfaces.

NOTE ADDED IN PROOF - The binding of type-3 fimbriae to type I collagen (Fig. 1) was due to unpure commercial collgen preparation. Highly purified type I collagen shows no reactivity with type-3 fimbriae (24).

REFERENCES

1. Sharon, N. 1986. in The lectrins. Properties, functions and applications in Biology and Medicine. Liener, I.E., N. Sharon and I.J. Goldstein, eds. pp. 493-526. Academic Press, Orlando.
2. Korhonen, T.K., R. Virkola, B. Westerlund, A.-M. Tarkkanen, K. Lahteenmaki, T. Sareneva, J. Parkkinen, P. Kuusela and H. Holthofer. 1988. Ant. Leueuwenhoek 54:411-420.
3. Korhonen, T.D., R. Virkola, B. Westerlund, H. Hothofer and J. Parkkinen. 1990. Curr. Top. Microbiol. Immunol. 151:115-127.
4. Virkola, R. B. Westerlund, H. Holthofer, J. Parkkinen, M. Kekokmaki and T.K. Korhonen. 1988. Infect. Immun. 56:2615-2622.
5. Nowicki, B., H. Holthofer, T. Sareneva, M. Rhen, V. Vaisanen-Rhen and T.K. Korhonen. 1986. Microb. Pathogen. 1:169-180.
6. Korhonen, T.K., J. Prkkinen, J. Hacker, J. Finne, A. Pere, M. Rhen and H. Holthofer. 1986. Infect. Immun. 54:322-327.

7. Korhonen, T.K., R. Virkola and H. Holthofer. 1986. Infeact. Immun. 54:328-332.
8. Virkola, R. 1987. FEMS Microbiol. Lett. 40:257-262.
9. Parkkinen, J. T.K. Korhonen, A.S. Pere, J. Hacker and S. Soinila. 1988. J. Clin. Invest. 81:860-865.
10. Parkkinen, J., R. Virkola and T.K. Korhonen. 1988. Infect. Immun. 56:2623-2630.
11. De Graaf, F.K., B.E. Krenn and P. Klaasen. 1984. Infect. Immun. 43:508-514.
12. Rhen, M., V. Vaisanen-Rhen, M. Saraste and T.K. Korhonen. 1986. Gene 49:351-3160.
13. Nowicki, B. C. Svanborg-Eden, R. Hull and S. Hull. 1989. Infect. Immun. 57:446-451.
14. Lindberg, F., B. Lund and S. Normark. 1986. Proc. Natl. Acad. Sci. USA 83:1891-1895.
15. Riegman, N., I. van Die, J. Leybusseb, W. Hoestra and H. Bergmans. 1988. Mol. Microbiol. 2:79-80.
16. Gerlach, C.-F., S. Clegg and B.L. Allen. 1989. J. Bacteriol. 171:1261-1270.
17. Speziale, P., M. Hook, T. Wadstrom and R. Timpi. 1982. FEBS Lett. 146:55-58.
18. Baloda, S.B., A. Faris, S. Froman and T. Wadstrom. 1987. FEMS Microbiol. Lett. 28:1-5.
19. Baloda, S.B., 1988. FEMS Microbiol. Lett. 49:483-488.
20. Westerlund, B., P. Kuusela, T. Vartio, I. van Die and T.K. Korhonen. 1989. FEBS Lett. 243:199-204.
21. Westerlund, B. P. Kuusela, J. Ristell, L. Risteli, T. Vartio, H. Rauvala, R. Virkola and T.K. Korhonen. 1989. Mol. Microbiol. 3:329-337.
22. Westerlund, B., J. Merenmies, H. Rauvala, A. Miettine, A.-K. Jarvinen, H. Hothofer and T.K. Korhonen. 1987. Microb. Pathogen. 3:117-127.
23. Nowicki, B., J. Moulds, R. Hull and S. Hull. 1988. Infect. Immun. 56:1057-1060.
24. Tarkkanen, A.-M., B.L. Allen, B. Westerlund, H. Holthofer, P. Kuusela, L. Risteli, S. Clegg and T.K. Korhonen. 1990. Mol. Microbiol. (in press).
25. Sareneva, T., H. Holthofer and T.K. Korhonen. 1990. Infect. Immun. (in press).
26. Timpl, R. 1986. Kidney Int, 30:293-298.
27. Kuehn, K. and R. Timpl, eds. 1982. New Trends in Basement Membrane Research, Raven Press, New York.
28. Vaisanen-Rhen, V. 1984. Infect. Immun. 46:401-407.
29. Korhonen, T.K., R. Virkola, V. Saisanen-Rhen and H. Holthofer. 1986. FEMS Microbiol. Lett. 35:762-766.
30. Gerlach, G.-F., B.L. Allen and S. Clegg. 1989. Infect. Immun. 57:219-224.
31. Old, D.C. and R.A. Adegbola. 1983. J. Med. Microbiol. 20:203-214.
32. Silverblatt, F.J. and I. Ofek. 1978. J. Inf. Dis. 138:664-667.
33. Miller, E.J. and S. Gay. 1987. Meth. Enzymol. 144:3-41.
34. Nowicki, B., J. Vuopio-Varkila, P. Viljanen, T.K. Korhonen and P.H. Makela. 1986. Microb. Parthogen. 1:335-347.
35. Van Die, I., E. Zuidweg, W. Hoekstra and H. Bergnams. 1986. Microbi. Pathogen. 1:51-56.
36. Parkkinen, J. and T.K. Korhonen. 1989. FEBS Lett., (in press).

37. Sodeinde, O.A. and J.D. Goguen. 1989. Infect, Immun, 57:1517-1523.
38. Liotta, L.A., C. Nageswaga Rao and U.M. Wewer. 1986. Ann. Rev. Biochem. 55:1037-1057.

P-FIMBRIAE: MOLECULAR ASPECTS OF THEIR STRUCTURE AND THEIR

APPLICATION AS CARRIERS OF FOREIGN ANTIGENIC DETERMINANTS

Irma van Die, Nico Riegman, Joost van Oosterhout,
Hans Bergmans, Anneke van der Zee, Hanneke Bes, and
Wiel Hoekstra

Department of Molecular Cell Biology, Padualaan 8
3584CH Utrecht, The Netherlands

INTRODUCTION

Urinary tract infections (UTI) are often caused by *Escherichia coli*. Properties associated with virulence include certain O- and K-antigens, P-fimbriae, production of hemolysin or colicin and resistance to serum bactericidal activity. Adherence of the bacteria to the uroepithelium is considered to be an important initial event in the pathogenesis of UTI. Adherence and the ability to cause mannose-resistant hemagglutination of human erythrocytes are associated in most UTI strains with the presence of P-fimbriae (1). P-fimbriae, or pap-pili, belong to a widely occurring family of fimbriae, filamentous surface components composed of a thousand protein subunits (for a review see 2). P-fimbriae specifically recognize the α-DGal(1-4)β-DGal moiety of glycolipids, which is also part of the P-blood group antigens (3,4). Among P-fimbriae a number of different serotypes are discriminated, designated as F7-F14 (5).

The gene clusters of a number of different P-fimbriae have been cloned (6-9), and their genetic organization has been studied in detail (10-12). Genetically and functionally, these gene clusters are very homologous, as several genes could be exchanged between the different gene clusters without influence on the expression of fimbriae or their adherence capacities (13).

MOLECULAR ANALYSIS OF THE COMPOSITION OF P-FIMBRIAE

P-fimbriae are complex filamentous structures, composed of several different protein subunits. They are encoded by a chromosomal gene cluster that, in the case studied in most detail, consists of 9 genes (12, see fig. 1). Four of these genes encode fimbrial structural elements, two genes are required for biogenesis, and two genes encode positive regulators of the gene cluster. Gene A codes for the main component of the fimbriae (approx. 95% of total fimbrial protein);

Microbial Surface Components and Toxins in Relation to Pathogenesis
Edited by E.Z. Ron and S. Rottem, Plenum Press, New York, 1991

23

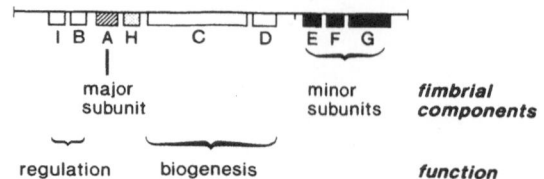

Fig. 1.   Genetic and functional organization of the P-fimbrial gene cluster (8,12). The boxes indicate genes
that encode proteins A-I.

genes E, F and G code for minor components, of which protein
G functions as the Gal-Gal specific adhesin (14). By immuno-
electronmicroscopy (15) we have studied the localization of
protein E and the adhesin protein G, using specific antisera
against these proteins. We have examined several types of P-
fimbriae, expressed by the uropathogenic strains AD110 (F7₁
and F7₂) and 21086 (F9) or recombinant _E. coli_ K12 strains
(F7₁, F7₂, and F11). The results (see figure 2) showed that
in most strains both proteins were found at the tip of the
fimbrial structure, as was also found by Lindberg et al (12)
for F13 (pap)fimbriae. By performing double-labeling experiments (figure 2) we found in all strains the adhesin exclusively in a complex with a number of protein E molecules; the
protein E "clump" however was also frequently observed without adhesin. Occasionally, these complexes were also found
along the fimbrial structure (which may be an artefact). The
F9 fimbriae of the wild type strain 21086 however clearly
carry proteinE-adhesin complexes all along the whole fimbrial
rod, similar to what has been described for type-1 fimbriae
(16-18). As expected, the F9 fimbriae of strain 21086 have a
relatively high content of protein G compared to the other P-
fimbrial types studied here, as was shown by SDS-PAGE analysis of isolated fimbriae (figure 3). Based on these results
we suppose that the localization of the proteinE/G complexes
in the fimbrial structure is determined by the relative
expression of G- and E protein with respect to the amount of
protein A. The way fimbriae are built may then be quite
uniform among different fimbrial types like P-fimbriae, type
1 fimbriae and probably also others, and depend on the relative expression of the various fimbrial components.

MOLECULAR ANALYSIS OF THE MAJOR FIMBRIAL SUBUNIT

The major subunit of the P-fimbriae was studied in
detail. The aminoacid sequences of the F7₁, F7₂, F11 and F13
fimbrial subunit were compared and a model has been presented
based on homology among the subunits (19). The N- and C-
termini of the various P-fimbrial subunits are very homologous; the region between the two cysteine residues in the
protein also is very homologous, although to a less extent.
The remaining region is more variable and here five hypervariable regions (HR) occur in which the aminoacids differ
completely. Recent sequence data (unpublished) of the F9
fimbrial subunit are in accordance with this model. It has
been shown that the major subunit of one gene cluster can be
assembled to fimbriae by the accessory proteins of other P-

24

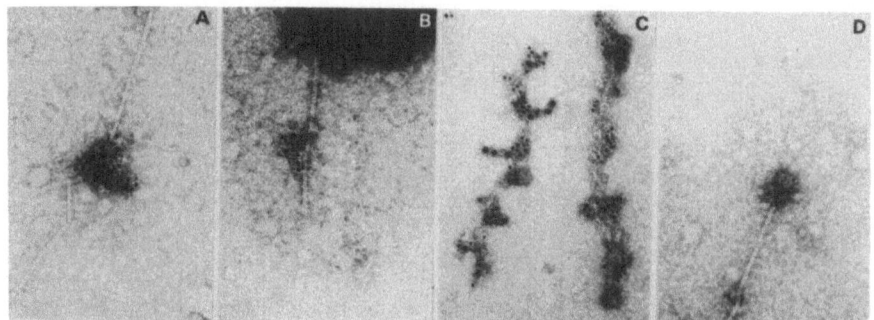

Fig. 2. Immunoelectron micrographs of fimbriae of: A, B and
D: HB101/pPIL110-75 (F7$_1$); C: uropathogenic *E. coli*
strain 21086 (F9). All preparations were treated
with 1:500 diluted polyclonal anti-E antiserum and
1:1000 diluted monoclonal antiserum Mab 6B10 (27),
that was kindly provided by Heinz Hoschutzky. The
11 nm gold particles represent the adhesin, the 6
nm gold particles represent protein E. HB101/pPIL-
110-37 (F7$_2$), HB101/pPIL291-15 (F11), and the uropathogenic *E. coli* strain AD110 (F7$_1$ and F7$_2$) show
a labeling pattern similar to the pattern observed
with HB101/pPIL110-75.

fimbrial gene clusters (13). This implies that the conserved
regions found in the major subunits most likely are important
for the biogenesis of the fimbriae. Given the serological
differences of P-fimbriae, the hypervariable regions are
thought to represent P-fimbrial epitopes. We have constructed
small mutations in hypervariable region 2 of the F7$_1$ subunit
and in region 1, 3 and 4 of the F11 subunit (20, unpublished
results). Mutant fimbriae were tested in an ELISA for their
binding capacity with specific monoclonal antibodies (table
1). The results show that mutations in these regions disturb
the binding by the fimbriae of one or more monoclonal antibodies, indicating that these regions represent (part) of an
epitope.

P-FIMBRIAE AS CARRIERS FOR FOREIGN ANTIGENIC DETERMINANTS

Fimbriae are produced as extracellular proteins in high
amounts, are immunogenic and easy to purify (21) and are
therefore potentially very attractive as carriers for foreign
antigenic determinants. The observation that hypervariable
regions are naturally present in the P-fimbrial subunit
suggests that these regions can be manipulated without loos-
ing the capacity to produce fimbriae. As the hypervariable
regions contain fimbrial epitopes, it might be expected that
foreign epitopes inserted at these places will be exposed in
the fimbrial structure.

In the F11 subunit unique cloning sites were constructed
in hypervariable regions 1 and 4. By site-directed mutagenesis of the F11 subunit gene, hypervariable region 4 was
removed and replaced by a 6-mer oligonucleotide with an *HpaI*
restriction site, resulting in plasmid pPIL291-1529(22). In

25

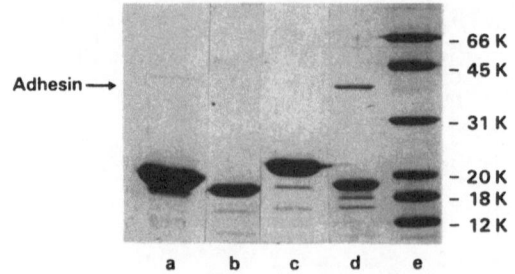

Fig. 3.   SDS-PAGE patterns of isolated fimbriae of [a] HB-
           101/pPIL110-37 (F7$_2$), [b] HB101/pPIL291-15 (F11),
           [c] HB101/pPIL110-75 (F7$_1$), [d] strain 21086 (F9),
           [e] molecular weight standard.

hypervariable region 1, 9 basepairs were replaced, such that
a *StuI* restriction site was created (plasmid pPIL291-15100,
see figure 4). As a model study, two oligonucleotides that
code for antigenic determinants of the vp1 coat protein of
foot-and-mouth disease virus (FMDV) were inserted in the
plasmids pPIL291-15100 and pPIL291-1529. The antigenic peptides used (see figure 5) are indistinguishable from FMDV in
its ability to bind the monoclonal antibodies (Mabs) MA11 and
MA18, which are neutralizing antibodies raised with FMDV type
A10 (23). The resulting recombinant strains were all able to
express hybrid F11/FMDV fimbriae, although with different

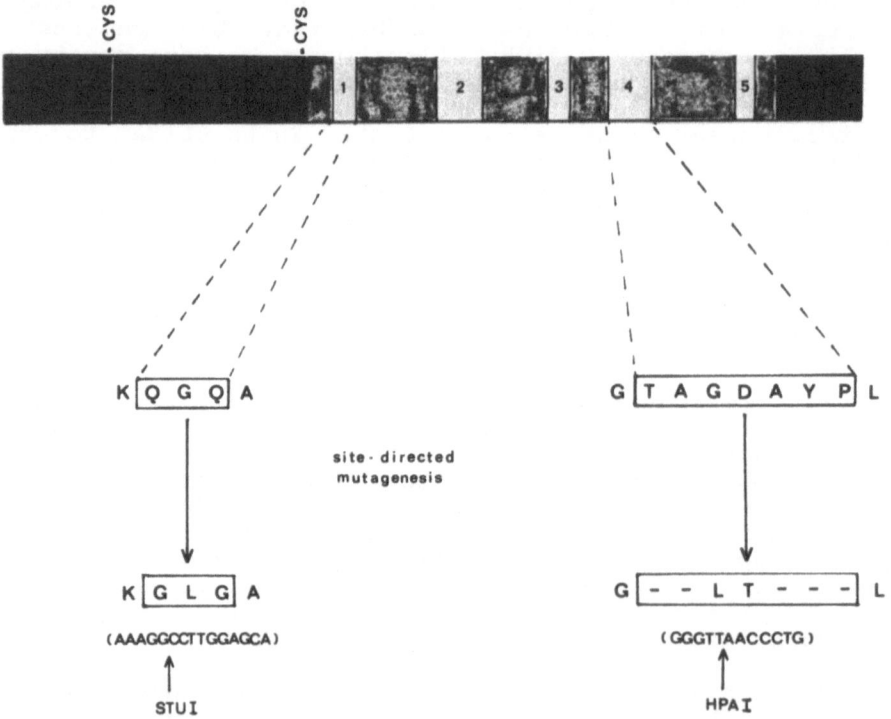

Fig. 4.   Model of P-fimbrial subunit with hypervariable
           regions (HR) 1-5, and construction of cloning sites
           in HR 1 and 4.

Table 1. Binding of F7₁ and F11 fimbriae and mutant derivatives of these fimbriae with F7₁ and F11 specific monoclonal antibodies (Mabs, see 26).

Characteristics of fimbriae	Binding with Mabs					
	M7-4	M7-6	M7-7	M7-13	M6-3	M6-6
	(F11 specificity)				(F7₁ specificity)	
wildtype F11	++	++	++	++	−	−
wildtype F7	−	−	−	−	++	++
mutations in						
HR 1 (F11)	+	−	++	++	−	−
HR 2 (F7₁)	−	−	−	−	++	−
HR 3 (F11)	++	++	−	−	−	−
HR 4 (F11)	++	++	−	+	−	−

Table 2. ELISA of different recombinant strains with F11 and FMDV specific monoclonal antibodies (Mabs, see 23,-26).

HB101 carrying plasmids	Insert in HR	FMDV epitope inserted	Reactivity with Mabs			
			M7-7 (F11)	M7-6 (F11)	MA11 (FMDV)	MA18 (FMDV)
pPIL291-15	−	−	+++	+++	−	−
pPIL291-15102	1	MA18	++	−	−	+++
pPIL291-15104	1	MA11	++	−	+++	−
pPIL291-1569	4	MA18	−	++	−	++
pPIL291-1567	4	MA11	−	+	+	−

Table 3. Expression of P-fimbriae and hybrid P/FMDV-fimbriae in *Salmonella* strains G30 and SHaroA (24,25).

Bacterial strains	Fimbriae	Binding of Mabs	
		M7-7 (F11)	MA18 (FMDV)
Salm. SHaroA/pPIL291-15	F11	+++	−
Salm. G30 /pPIL291-15	F11	++	−
HB101 /pPIL291-15	F11	+++	−
Salm. SHaroA/pPIL291-15102	F11/FMDV	++	++
Salm. G30 /pPIL291-15102	F11/FMDV	+	+
HB101 /pPIL291-15102	F11/FMDV	++	++

A. MA18-epitope

A TAT AAA CAG AAG ATC ATC GCC CCG GG
Tyr Lys Gln Lys Ile Ile Ala Pro

B. MA11-epitope

A TCG CGC CGT GGT GAT CTG GGA TCC TTA GCA CCA CGT GTT AA
Ser Arg Arg Gly Asp Leu Gly Ser Leu Ala Pro Arg Val

Fig. 5.   Oligonucleotides coding for FMDV antigenic peptides
that bind the FMDV specific Mabs MA11 and MA1 (23).

efficiencies. The strains were tested in an ELISA for their
binding capacities with both F11 and FMDV specific antisera.
As the results show (table 2), insertion of the MA18 epitope
in HR1 or HR4, and of the MA11 epitope in HR1 results in an
efficient, although slightly reduced fimbriae production.
Insertion of the MA11 epitope in HR4 results in a low fimbriae production.

The antigenicity of both FMDV epitopes in the hybrid fimbriae is much better in HR1 than in HR4. These results were
confirmed by immuno-electron microscopy (e.g. figure 6). The
differences found between HR1 and HR4 suggest that HR1 is
more suitable for insertion of antigenic peptides than HR4.
Preliminary data however with other antigenic peptides showed
that some of these peptides were more successfully inserted
in HR4, both with respect of fimbriae production and
antigenicity.

EXPRESSION OF (HYBRID) P-FIMBRIAE IN SALMONELLA TYPHIMURIUM

A possible application of hybrid P-fimbriae carrying
foreign antigenic determinants is to deliver the fimbriae in
a suitable host such as *Salmonella typhimurium* to the immune
system via the oral route. Oral vaccination is specifically
important when mucosal immunity is desired. Attenuated *Salmonella* strains have been shown to be very suitable for this
purpose (24).

To test whether the development of live oral vaccines can
be envisaged, expression of P-fimbrial gene clusters in two
attenuated *Salmonella typhimurium* strains, G30 and SHaroA
(24,25), was studied. The results in table 3 show that expression of both the hybrid FMDV/F11 and the F11 wild type
fimbriae is as efficient in the SHaroA strain as in *E. coli*
K12. Expression in the G30 strain is reduced, but still
considerable. Also the viral epitope was recognized efficiently by MA18 in the *Salmonella* strains.

CONCLUSIONS AND PERSPECTIVES

We have demonstrated that the hypervariable regions of *E.
coli* P-fimbriae are suitable places to insert foreign antigenic determinants. The foreign epitopes are antigenic in the
fimbriae. The availability of several regions for insertion
in the subunit allows to select the most suitable place for

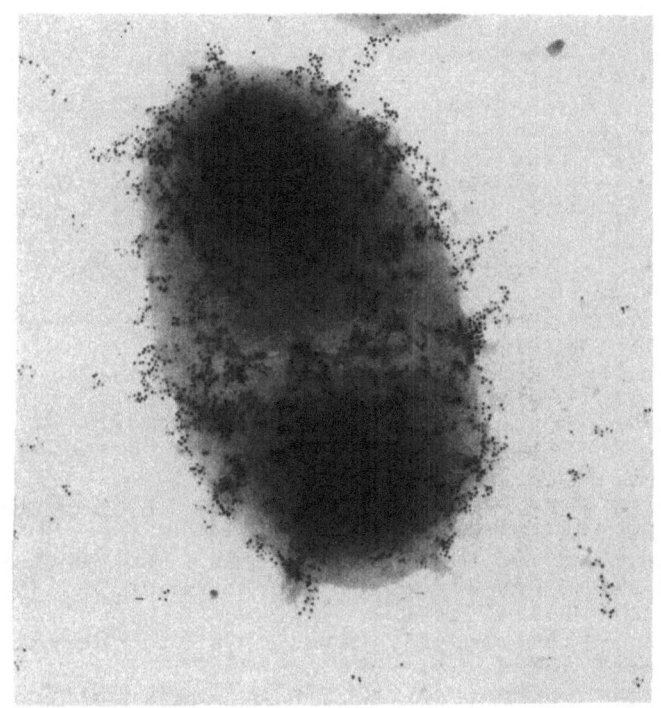

Fig. 6.   Immunoelectron micrograph of HB101/pPIL291-15104,
           treated with the FMDV specific Mab MA11 (23), in a
           dilution of 1:1000, and labeled with 7 nm proteinA-
           gold complexes. A similar labeling pattern was ob-
           served with HB101/pPIL291-15102 with MA18, while
           HB101/pPIL291-1569 and HB101/pPIL291-1567 showed a
           very poor labeling with these Mabs.

a specific determinant. It also opens the possibility to
insert more than one determinant in the fimbriae. Hybrid
fimbriae, as described here, might be valuable tools in the
development of serodiagnostics. If immunogenic (a question
that is currently under investigation) hybrid fimbriae might
be suitable as components of a subunit vaccine. Alternatively, recombinant plasmids coding for expression of hybrid
fimbriae can be introduced in *Salmonella* strains, and the
recombinant strains used as an oral vaccine. Expression of
F11 wild type fimbriae as well as pPIL291-15102 encoded
fimbriae in the attenuated *Salmonella* G30 and SHaroA strains
was found to be efficient. The presence of extracellular
surface proteins on *Salmonella* seems very attractive when
developing live oral vaccines.

REFERENCES

1. Vaisanen, V., Elo, J., Tallgren, L.G., Siitonen, A., M-
    akela, P.H., Svanborg-Eden, C., Kallenius, G., Svenson,
    S.B., Hultberg, H., and Korhonen, T.K. 1981. Lancet ii:
    1366-1369.
2. Klemm, P. 1985. Rev. Infect. Dis. 7:321-340.
3. Kallenius, G., Svensson, S.B., Mollby, R., Cedergren, B.,

Hultberg, H., and Winberg, J. 1981. Lancet ii:604-606.

4. Leffler, H. and Svanborg-Eden, C. 1980. FEMS Microbiol. Lett. 8:127-134.

5. Orskov, I. and Orskov, F. 1985. J. Hyg.95:551-575.

6. Hull, R.A., Gill, R.E., Hsu, P., Minshew, B.H., and Falkow, S. 1981. Infect. Immun. 33:933-938.

7. Van Die, I., Van den Hondel, C., Hamstra, H.J., Hoekstra, W. and Bergmans, H. 1983. FEMS Microbiol. Lett. 19:77-82.

8. Van Die, I., Spierings, G., Van Megen, I., Zuidweg, E., Hoekstra, W. and Bergmans, H. 1985. FEMS Microbiol. Lett. 28:329-334.

9. De Ree, J.M., Schwillens, P. and Van den Bosch, J.F. 1985. FEMS Microbiol. Lett. 29:91-97.

10. Norgren, M., Normark, S., Lark, D., O'Hanley, P., Schoolnik, G., Falkow, S., Svanborg-Eden, C., Baga, M., and Uhlin, B.E. 1984. EMBO J. 3:1159-1169.

11. Van Die, I., Van Megen, I., Hoekstra, W. and Bergmans, H. 1984. Mol. Gen. Genet. 194:528-533.

12. Lindberg, F.P., Lund, B., Johansson, L. and Normark, S. 1987. Nature London. 328:84-87.

13. Van Die, I., Van Megen, I., Zuidweg, E., Hoekstra, W., De Ree, H., Van den Bosch, H. and Bergmans, H. 1986. J. Bacteriol. 167:407-410.

14. Lund, B., Lindberg, F., Marklund, B.I., Normark, S. 1987. Proc. Natl. Acad. Sci. USA 84:5898-5902.

15. Van Bergen en Henegouwen, P.M.P. and Leunissen, J.L.M. 1986. Histochem. 85:81-87.

16. Abraham, S.N., Goguen, J.D., Sun, D., Klemm, P., and Beachey, E.H. 1987. J. Bacteriol. 169:5530-5536.

17. Abraham,S.N., Goguen, J.D.,and Beachey, E.H. 1988. Infect. Immun. 56:1023-1029.

18. Krogfelt, K.A. and Klemm, P. 1988. Microb. Pathogen. 4: 231-238.

19. Van Die, I., Hoekstra, W., and Bergmans, H. 1987. Microb. Pathogen. 3:149-154.

20. Van Die, I., Riegman, N., Gaykema, O., Van Megen, I., Hoekstra, W.,Bergmans, H., De Ree, H. and Van den Bosch, H. 1988. FEMS Microbiol. Lett. 49:95-100.

21. Korhonen, T.K., and Rhen, M. 1982. Ann. Clin. Res. 14:272-277.

22. Van Die, I., Wauben, M., Van Megen, I., Bergmans, H., Riegman, N., Hoekstra, W., Pouwels, P., and Enger-Valk, B. 1988. J. Bacteriol. 170:5870-5876.

23. Meloen, R.H, Puijk, W.C., Meijer, D.J.A., Lankhof, H., Posthumus, W.P.A., and Schaaper, W.M.M. 1987. J. Gen. Virol. 68:305-314.

24. Dougan, G., Hormaeche, C.E., Maskell, D.J. 1987. Parasite Immunol. 9:151-160.

25. Stevenson, G. and Manning, P. 1985. FEMS Microbiol. Lett. 28:317-321.

26. De Ree, J.M., Schwillens, P., and Van den Bosch, J.F. 1985. Infect. Immun. 50:900-904.

27. Hoschutzky, H., Lottspeich, F. and Jann, K. 1989. Infect. Immun. 57:76-81.

ENTEROAGGREGATIVE *ESCHERICHIA COLI,* A NEW DIARRHEAL AGENT

Myron M. Levine, Bernadette Baudry,
Stephen Savarino, Pablo Vial, James Kaper

Center for Vaccine Development, Department of Medicine
University of Maryland School of Medicine, and The
Medical Biotechnology Center University of Maryland, 10
S. Pine St., Baltimore, Maryland 21201, U.S.A.

INTRODUCTION

The ability of *Escherichia coli* strains to adhere to HEp-2 cells in tissue culture correlates with their capacity to cause diarrhea. Three distinct patterns of adherence have been described: localized, diffuse and aggregative (1). Localized adherence (LA) is characteristic of enteropathogenic *E. coli* of classical infant diarrhea O:H serotypes (2). This property requires the presence of EPEC Adherence Factor (EAF) genes which are located on a plasmid (2,4,5). Several recent epidemiological studies have clearly demonstrated that *E. coli* isolates exhibiting LA are isolated significantly more often from infants and young children with diarrhea than from matched controls (3,6,7); the vast majority of such isolates fall into classical EPEC O serogroups (3). Diffuse adherence *E. coli* (DAEC) are probably the least studies and understood of the HEp-2 cell adherent *E. coli*. In the one strain extensively studied, diffuse adherence requires the expression of antigenically distinct fimbriae, the structural the expression of antigenically distinct fimbriae, the structural genes of which are chromosomal (8). There is controversy over whether of not DAEC are diarrheagenic (1,3,6,7,9,10).

The epidemiological significance of EAgEC, their microbiological and genetic characteristics and their mechanisms of pathogenesis ar currently being explored. The current state of knowledge on this subject will be summarized in this review.

ARE EAggEC DIARRHEAL PATHOGENS?

The ability of a suspect pathogen to cause diarrhea in humans can be suspected or verified by four sources of information:

Microbial Surface Components and Toxins in Relation to Pathogenesis
Edited by E.Z. Ron and S. Rottem, Plenum Press, New York, 1991

31

1) Carefully-performed epidemiologic studies that compare the rate of isolation of the suspect pathogen from cases of diarrhea versus from appropriately matched controls.

2) Volunteer studies in which a prototype strain is fed to volunteers under controlled conditions where intense clinical surveillance can be maintained and clinical specimens can be bacteriologically cultured and examined.

3) Epidemiological investigation of an outbreak that incriminates the putative pathogen.

4) Clear-cut evidence of pathogenicity in a relevant animal model. The evidence that incriminates EAggEC as an etiologic agent of diarrhea will be discussed below.

EPIDEMIOLOGICAL STUDIES

Several epidemiological case/control studies have been carried out so far in widely separated geographic regions of the world wherein EAggEC have been sought. The first suggestion that these *E. coli* may be diarrheagenic agents came from a study of summer diarrhea in infants and young children in Santiago, Chile (1,3). In this study EAggEC were isolated significantly more often from cases (36%) than controls (23%) (p=0.04). Bhan et al (11) looked for EAggEC in a cohot of children less than five years of age in rural India who were prospectively studied for diarrheal disease. There was no difference in the isolation of EAggEC between controls (9.9%) and children with acute diarrhea (s14 days duration). However, the isolation from children with persistent diarrhea (>14 days duration) (29.5%) was significantly greater than controls (p=0.0006) or cases of acute diarrhea (p=0.0052). In a subsequent hospital-based case/control study of the etiology of persistent diarrhea, Bhan et al (14) have again found EAggEC to be isolated significantly more often from cases than controls. In contrast, Gomes et al (6), found EAggEC in equal frequency in 100 cases of infant diarrhea (10%) and 100 matched controls (8%) in Sao Paulo, Brazil.

VOLUNTEER STUDIES

Strain 221 was isolated from a U.S. traveler to Mexico with diarrhea (10). Mathewson et al (10) originally described strain 221 as being of non-EPEC O serogroup and exhibiting localized adherence. However, subsequently, Vival et al (13) and Levine et al (3) demonstrated that strain 221 is a typical EAggEC isolated that manifests the aggregative pattern in the HEp-2 cell assay and hybridizes with the EAggEC probe. Mathewson et al (10) fed strain 221 to volunteers at a dose of 10^8 or 10^{10} viable organisms. Diarrhea occurred in five of 16 volunteers.

OUTBREAK OF DIARRHEA

Therefore, there has been no outbreak of diarrhea where EAggEC have been clearly identified as the causative agent. However, in the 1970s an outbreak of diarrhea in pediatric intensive care units in two London hospitals led to the identification of several unusual *E. coli* strains from cases (14). Several of these isolates expressed a fimbrial hamagglutinin (14) antigenically distinct from known fimbrial colonization factors. We have now shown that several of these same isolates are EAggEC, i.e., they manifested the aggregative pattern in the HEp-2 assay and hybridize with a DNA probe that is highly specific for EAggEC.

ANIMAL MODELS

Vial et al (13) inoculated isolated intestinal loops of rabbits and rats with EAggEC strains; other loops were inoculated with EPEC strain E2348/69 as a positive control and non-pathogenic strain HB101 as a negative control for histopathological studies. All four EAggEC strains tested in rabbit and rat intestinal loops induced striking histopathological changes. These included marked shortening of the villi, hemorrhagic necrosis of the villus tips, and a mild inflammatory reaction with edema and mononuclear infiltration of the submucosa. This pathological picture, which is distinct from that seen with any other category of diarrheagenic *E. coli*, is compatible with the effect of a potent cytotoxin.

Several strains of EAggEC have been fed to gnotobiotic piglets by Saul Tzipori and coworkers at the Royal Children's Hospital, Melbourne in a collaboration with investigators from the Center for Vaccine Development of the University of Maryland (Tzipori et al, manuscript in preparation). Two of the isolates caused overt diarrhea. Pathological examination of the intestine revealed EAggEC manifesting aggregative phenomenon *in vivo*. The gnotobiotic piglet model is a well-established system that has proved very useful for studying the pathogenesis of other bacterial enteropathoens of humans, including EPEC (15), enterohemorrhagic *E. coli* (16) and *Yerisinia enterocolitica* (17).

In summary, evidence from several sources suggests that EAggEC are diarrheal pathogens.

CHARACTERIZATION OF EAggEC

O:H Serotypes

The other categories of diarrheagenic *Escherichia coli*, including enterotoxigenic, enteropathogenic, enteroinvasive and enterohemorrhagic, fall into distinct sets of O:H serotypes (18). Therefore, attempts were made to serotype EAggEC isolates and to determine if they fell into the serotypes characteristic of any of the other categories of diarrheagenic *E. coli*. In these early attempts, many of the EAggEC initially appeared to be serologically rough in that they autoagglutinated in saline (13). Since it was considered unlikely that true enteric pathogens would lack smooth O antigens, further studies were carried out. It was found that the "rough-appearing" strains were not recognized by rough specific phages. Moreover, it was possible to extract smooth lipopolysacchride O antigens from the strains by the hot water-phenol method and visualize them by polyacrylamide gel electrophoresis. Common EAggEC O serogroups that have been identified are 03, 015 and 089, while H33 and H2 are common H types. These O:H serotypes are not characteristic of other categories of diarrheagenic *E. coli*.

Plasmids

Virtually all EAggEC harbor a plasmid circa 60 Md in size which is necessary for EAggEC to manifest the aggregative phenomenon (13). Transfer of this plasmid to *E. coli* HB101 was accompanied by transfer of the aggregative phenomenon (13). Notably, with two strains plasmid transfer was also accompanied by transfer of the ability to express smooth O lipopolysaccharide.

Vival et al (13) showed that plasmids of EAggEC share considerable homology.

Fimbriae

A group of EAggEC strains that were grown on solid CFA agar at 37°C; the strains that were non-hemagglutinating when tested with guinea pig erythrocytes (i.e. they were not expressing type 1 somatic fimbriae), were examined by electron microscopy. Most strains revealed the presence of rigid fimbriae 6-7 nm in diameter (13). A monospecific antiserum was prepared against the purified fimbriae of a prototype EAggEC strain, 17-2. This antiserum reacted with colony blots made from a number of other EAggEC strains grown at 37°C but not with the same strains grown at 18°C.

IDENTIFICATION OF EAggEC

HEp-2 Cell Assay

The classic method for identifying EAggEC is to culture test organisms overnight in L broth and then inoculate monolayers of HEp-2 cells in tissue culture with the bacterial cultures. The method recommended is that of Nataro et al (1) which is essentially the 3-hour HEp-2 assay described by Cravioto et al (19). The monolayer should be 25-50% confluent. This assay is readily applicable to the testing of small numbers of strains but is not practical for large-scale epidemiological studies wherein large numbers of isolates must be tested.

A 1.0 kb fragment of plasmid DNA derived from the 60 Md plasmid of a Peruvian EAggEC strain hybridized under stringent conditions with colony blot DNA of 20 of 41 EAggEC strains (49%) but not with any of 40 enterotoxigenic, enteropathogenic, enterohemorrhagic of enteroinvasive *E. coli* strains (13). These observations suggested that it might be possible to develop a DNA probe to identify EAggEC; DNA probes are tools that are highly suited to the testing of large numbers of isolates (3). Pursuing this approach, Baudry et al (B. Baudry, S. Savarino, P. Vial, M. Levine, submitted for publication) have isolated a 1.0 kb fragment of DNA from the 60 Md plasmid of a prototype EAggEC strain that shows 89%

sensitivity and 99% specificity in identifying EAggEC when tested against several hundred well-characterized isolates (using the HEp-2 assay as the "gold standard").

SUMMARY

Strains of *E. coli* that exhibit the aggregative pattern in the HEp-2 cell assay represent a distinct category of *E. coli* and evidence from several sources suggests that these organisms are diarrheal pathogens, particularly as a cause of persistent diarrhea. The aggregative property is dependent on the presence of a circa 60 Md plasmid that is also associated with the expression of novel fimbriae ad smooth O antigens. The drawbacks of the cumbersome HEp-2 cell assay as the method of identifying EAggEC have been overcome with the development of a sensitive and specific DNA probe. The probe, which consists of a 1 Kb fragment of DNA derived from the plasmid of a prototype EAggEC strain, is amenable to the testing of large numbers of isolates and thus should expedite the performance of epidemiological studies.

REFERENCES

1. Nataro J.P., J.B. Kaper, R. Robins-Browne, V. Prado, V. Vial, M.M. Levine. 1987. Pediat. Infect. Dis. J. 6:829-831.
2. Nataro J.P., I.C. Scaletsky, J.B. Kaper, M.M. Levine, L.R. Trabulsi. 1985. Infect. immun. 48:378-383.
3. Levine M.M., V. Prado, R.M. Robins-Browne, H. Lior, J.B. Kaper, S. Moseley, K. Gicquelais, J.P. Nataro, P. Vial and B. Tall. 1988. J. Infect. Dis. 158:224-228.
4. Baldini M.M., J.B. Kaper, M.M. Levine, C.D.A. Candy and H.W. Moon. 1983. J Pediat. Gastroenterol Nutr. 2:534-538.
5. Nataro J.P., M.M. Baldini, J.B. Kaper, R.E. Black, N. Bravo and M.M. Levine. 1985. J. Infect. Dis. 152: 560-5.
6. Gomes T.A., P.A. Blake and L.R. Trabulsi. 1989. J Clin. Microbiol 1989; 27:266-269.
7. Cravioto A., R.E. Reyes, R. Ortega, G. Fernandez, R. Hernandez, D. Lopez. 1988. Epidemiol. Infect. 101:123-134.
8. Moseley, S., C.R. Clausen and A.L. Smith. 1985. Abstract 1128. In: Program and abstracts of the 25th Interscience Conference on Antimicrobial Agents and Chemotherapy. Washington, D.C.: American Society for Microbiology.
9. Mathewson J.J., R.A. Oberhelman, H.L. DuPont, F.J. De la Cabada and E.V. Garibay. 1987. J. Clin. Microbiol. 25: 1917-1919.
10. Bhan M.K., P. Raj, M.M. Levine, J.B. Kaper, N. Bhandari, R. Srivastava, R. Kumar, and S. Sazawal. 1989. J. Infect. Dis. (in press).
11. Bhan M.K., personal communication.
12. Vial P., R.M. Robins-Browne, H. Lior, V Prado, J.B. Kaper, A. Elsayed, and M.M. Levine. 1988. J. Infect. Dis. 158: 70-79.
13. Candy D.C.A., T.S.M. Leung, R.H.K. Mak, J.T. Harries, W.C. Marshall, A.D. Phillips, R.M. Robins-Browne, M.V.

Chadwick and M.M. Levine. 1982. FEMS Microbiol. Lett 15:325-329.

14. Tzipori, S., R.M. Robins-Browne, G. Gonis, J. Hayes, M. Withers and E. McCartney. 1985. Gut 26:570-578.

15. Tzipori, Ss, H. Karch, I.K. Wachsmuth, R.M. Robins-Browne, A.D. O'Brien, H. Lior, M.I. Cohen, J. Smithers and M.M. Levine. 1987. Infect. Immun. 50:3117-3125.

16. Robins-Browne R.M., S. Tzipori, G. Gonis, J. Hayes, M. Withers and J.K. Prpic. 1985. J. Med. Microbiol. 19: 197-308.

17. Levine, M.M. 1987. J. Infect. Dis. 155:377-389.

18. Cravioto, A., R.J. Gross, S.M. Scotland and B. Rowe. 1979. Curr. Microbiol. 3:95-99.

FIMBRIAE OF HUMAN ENTEROTOXIGENIC *ESCHERICHIA*

COLI AND CONTROL OF THEIR EXPRESSION

Cyril J. Smyth, Maire Boylan[1],
Helen M. Matthews and David C. Coleman

Department of Microbiology, Moyne Institute, Trinity
College, Dublin, Republic of Ireland and [1]Department of
Biochemistry, Emory University, School of Medicine
Atlanta, Georgia 30322, U.S.A.

INTRODUCTION

One of the major causes of diarrhoea in infants in
developing countries and in travellers to the Third World is
enterotoxigenic *Escherichia coli* (ETEC) (1-4). Infection is
acquired by ingesting contaminated food or water. Two virulence attributes of ETEC are essential for pathogenicity,
namely enterotoxin production and the ability to adhere to
and to colonize the epithelium of the small intestine. Bacterial growth *in situ* and elaboration of enterotoxin(s) leads
to increased water and electrolyte secretion into the intestinal lumen and thus the diarrhoeic response and dehydration.
Individual strains of ETEC may produce either heat- labile
enterotoxin (LT) or heat-stable enterotoxin (ST), or both.
The heat-labile toxin resembles the enterotoxin of *Vibrio
cholerae* in molecular weight, serological properties and mode
of action (5,6). The heat-stable toxin is a polypeptide of
low molecular weight which is poorly antigenic (6). Two types
of ST have been identified, STa or ST1 being produced by
human ETEC strains. LT and ST activate adenylate cyclase and
guanylate cyclase, respectively (6).

ADHESINS OF HUMAN ETEC

Adhesins are macromolecular structures present on the
surface of bacteria which mediate attachment to the host cell
surface. Adherence of ETEC to the brush borders of epithelial
cells of the proximal small intestine is a crucial event in
their pathogenesis. A clear correlation has been established
between the presence of specific adhesins on the surface of
bacteria and their ability to adhere to a given host mucosal
surface (7-10). These adhesins allow the bacteria to overcome
the clearing mechanisms that protect the epithelial surface
(e.g. peristalsis, mucus) and also provide a structure which
binds to a distinct receptor on the mucosa or in the glyco-
calyx of epithelial cells (11-14).

Microbial Surface Components and Toxins in Relation to Pathogenesis
Edited by E.Z. Ron and S. Rottem, Plenum Press, New York, 1991

37

A special class of adhesins are the proteinaceous non-flagellar appendages called fimbriae or pili which emanate from the bacterial surface. Each type of fimbria is composed of a major structural subunit protein called fimbrin or pilin. Additional minor protein subunits have also been identified (e.g. 15-17). In some cases adhesion has been shown to be mediated by minor fimbriae-associated proteins and not by the major structural subunit of the fimbriae (15,16,18-20), although such structural differentiation has not, as yet, been demonstrated for fimbriae on human ETEC.

The nomenclature of fimbriae on human ETEC is confusing at present. Terms such as colonization factor antigen (CFA), coli-surface antigen (CS antigen) and putative colonization factor (PCF) reflect their mode of detection or presumed role when first discovered. Some of these fimbrial types found on human ETEC appear to be present on only a relatively small number of serovars or serotypes (Table 1), e.g. CS1 on strains of serotype O6:K15:H16 or H-, PCFO159:H4 on strains of serotype O159:H4 or H20, whereas others occur naturally on strains belonging to a range of O-serovars, e.g. CFA/I, CS3, CS6 and CS5/CS6. The reasons for this distribution of fimbrial occurrence in wild-type isolates are still unclear. Moreover, ETEC expressing these different adhesins possess characteristic enterotoxin-producing phenotypes and give rise to mannose-resistant haemagglutination (MRHA) of characteristic patterns (Table 1). The different haemagglutination patterns reflect different receptor specificities on the red blood cells for the different fimbrial ligands, although little is known about the nature of these receptors (25, 38-40).

Serological surveys of ETEC of human origin for these fimbrial types have also revealed that there are still some O-serovars in which colonization factors or haemagglutinins or tissue adhesins have not, as yet, been identified. Many of these ETEC produce only LT and belong to serovars that occur infrequently (Table 1), which has led to less intense study on what are currently regarded as a pot-pourri of strains of minor importance. Should mass vaccination against ETEC, based on their adhesins, be employed in developing countries at a future date, these minor strains might become the predominant ETEC flora in surveys of diarrhoea in immunized subjects. Such shifts in the prevalent ETEC flora have already been seen in animals vaccinated with adhesins such as the K88 and K99 antigens (41,43).

MORPHOLOGY AND STRUCTURE OF HUMAN ETEC FIMBRIAE

Morphologically, the fimbriae on human ETEC strains can be divided into two types:

(i)   Those resembling type 1 fimbriae (43) which are visualized under the electron microscope as thin, rigid rod-like structures 6-7 nm in diam. (Table 2). Brinton (43) proposed that the subunits of type 1 fimbriae with a molecular mass of 17,000 daltons were arranged in a right-handed helix with a pitch distance of 6.32 nm, 3.125 subunits per turn and an axial pore of 2 nm diam. such that the supramolecular arrangement gave a rigid rod-shaped structure. It is probable,

Table 1. Fimbriae (adhesins) of human ETEC in relation to O serovar, enterotoxin type and mannose-resistant haemagglutination.

Fimbriae (adhesins)*	O Serovars or O:K:H Serotypes	Enterotoxin type	MRHA of indicated species	References
CFA/I	O4, O7, O20, O25, O63, O78, O110, O126, O128, O136, O153, O159	ST^+ or ST^+LT^+	Human, Chicken, Bovine	Evans et al.(21); McConnell et al.(22)
CFA/II:				Evans & Evans (23)
CS1/CS3	O6:K15:H16 or H−, O139:H28	ST^+LT^+	Bovine, Chicken**	Cravioto et al.(24);
CS2/CS3	O6:K15:H16 or H−	ST^+LT^+	Bovine, Chicken**	Smyth (25,26);
CS3 only	O8, O9, O78, O80, O85, O115, O128, O139, O168	ST^+LT^+	Bovine	McConnell et al.(22)
CFA/III−CS6	O25:H16 or H−	LT^+	−	Honda et al.(27,28) McConnell & Rowe(29) McConnell et al.(22)
CFA/IV:***				
CS4/CS6	O25:H42	ST^+LT^+	Human, Bovine	Thomas & Rowe (30)
CS5/CS6	O6, O29, O92, O114, O115, O167	ST^+	Human, Bovine	Thomas et al. (31)
CS5 only	O15, O114	ST^+LT^+ or LT^-	Human, Bovine Guinea-pig	
CS6 only	O27, O92, O148, O153, O159, O169	ST^+		McConnell et al. (22,33)
PCFO159:H4	O159:H4 or H20	ST^+LT^+		Tacket et al. (32)
PCFO166	O166:H27 or H48	ST^+	Human, Bovine	McConnell et al.(22)
CFO148:H28	O148:H28	ST^+LT^+	?	Knutton et al.(34)
CS17	O8, O15, O114	LT^+		McConnell, M.M. (personal commun.)
None of above	O2, O9, O15, O17, O20, O21, O40, O41, O48, O59, O64, O79 O80, O88, O101, O104, O112 O114, O117, O146, O159	ST^+	?	Levine (35); McConnell et al.(22) McConnell & Rowe(29)
None of above	O7, O9, O15, O17, O20, O21, O41, O41, O48, O59, O64, O79, O80, O88, O101, O104, O112, O114, O117, O146, O159	LT^+	?	Levine (35); McConnell et al.(22) McConnell & Rowe(29)
None of above	O1, O7, O17, O20, O60, O98, O109, O159	ST^+LT^+	?	Levine (35); McConnell et al.(22) McConnell & Rowe(29)

*CFA = Colonization factor antigen; CS = Coli surface-associated antigen;
 PCF = Putative colonization factor.
**Human erythrocytes from freshly drawn blood samples from black donors are agglutinable by bacteria expressing CS1 and CS2 fimbriae (37,24).
***Previously called E8775 and PCF8775 (30,31); The morphological nature of the CS6 antigen has not, as yet, been identified (36)

Table 2. Properties of fimbriae of human ETEC.

Fimbriae (adhesins)	Morphological type of fimbria*	Molecular Weight of subunit	N-terminal sequence	References
CFA/I	R	13,000	–	Willshaw et al.(51); Smyth (25)
		15,000	+	Klemm (45)
CFA/II:				
CS1	R	16,900	+	Matthews (47)
		16,300	–	Smyth (25)
CS2	R	17,000	+	Klemm et al. (48)
		16,500	+	Matthews (47)
		16,000	–	Sjoberg et al. (38)
		15,800	+	Smyth (25)
CS3	W	16,000	+	Boylan et al. (49)
		14,800	–	Smyth (25)
		14,500;15,500**	–	Levine et al. (52)
CFA/III	R	18,000	–	Honda et al. (28)
CFA/IV:				
CS4	R	22,000	–	Wolf et al. (36)
		17,000	+	McConnell et al. (53)
CS5	R/W	23,000	+	Heuzenroeder et al. (50)
		21,000	–	McConnell et al. (53)
CS6	?	14,500;16,000***	–	Wolf et al. (36)
CF0148:H28	W	?	–	Knutton et al. (34)
PCF0159:H4	R	19,000	–	Tacket et al. (32)
PCF0166	R	15,500;17,000**	–	McConnell et al. (22)
CS17	R	?	–	McConnell, M.M.(personal communication)

*R = rigid, rod-shaped, 6-7 nm. diam.; W = wiry, flexible, 2-3 nm. diam.

**Two distinct bands revealed by SDS-PAGE

***Strains of serotypes 025:H42, 027:H7 and 027:H20 revealed one band of 14,500 daltons whereas strains of serovers 0148, 0159 and 0167 revealed two bands by SDS-PAGE.

but not proven by X-ray diffraction and crystallography, that the morphologically identical fimbriae on human ETEC also comprise helices of subunits, but with different pitch distances and numbers of subunits per turn of the helix depending on the molecular weights of the subunits.

(ii) Thin wiry, wavy or curly structures with a diam. of 2 to 4 nm resembling the K88 and F41 fimbriae found on porcine and bovine isolates of ETEC, respectively, e.g. CS3 fimbriae. No X-ray diffraction or crystallographic studies have been carried out on these flexible fibrils. It is thought that this type of fimbria also comprises structural subunits in a helical arrangement, but that the number of subunits per turn of the helix results in an open-structured helix rather than a closed one around an axial pore (7). The open helical structure proposed for *Bordetella pertussis* fimbriae with a pitch distance of 6.5 nm and 2.5 repeating units per turn may be useful as a model for these flexible, wiry fimbriae of human ETEC (44).

Whereas some of these fimbriae have only ever been detected by themselves on strains of human ETEC, e.g. CFA/I, PCFO166 and PCFO159:H4 fimbriae, other fimbriae (adhesins) occur more commonly in specific combinations, e.g. CFA/III fimbriae + CS6 antigen, CS4 fimbriae + CS6 antigen, CS5 fimbriae + CS6 antigen, and CS1 or CS2 fimbriae + CS3 fimbriae. The presence of more than one fimbrial gene cluster in a strain raises the possibility of interaction between gene clusters in the regulation of expression or in the biogenesis of fimbriae on the bacterial cell surface, although investigations to date have not revealed this scenario.

Many of these fimbriae have now been purified and the molecular weights of their structural subunits have been characterized (Table 2). The N-terminal amino acid sequences of several of these fimbrial subunits have been determined using purified fimbriae or have been deduced from the nucleotide sequences of cloned genes encoding structural subunits, or both (Table 3). There is a substantial degree of homology between the N-terminal amino acid sequences of CS1, CS2, CS4 and CFA/I fimbrial subunits. In contrast, the N-terminal amino acid sequences of the subunits of CS3 and CS5 fimbriae show no close homology either with each other or with those of the CFA/I, CS1, CS2 and CS4 fimbrial subunits (Table 3).

The N-terminal amino acid sequence of fimbrial subunits is a region which has been proposed to be important in their polymerization (7,8,10). This may account for the significant homology in amino acid sequence that exists between the N-termini of the CFA/I, CS1, CS2 and CS4 fimbrial subunits. The N-terminal regions of CFA/I, CS1, CS2 and CS4 fimbrial subunits also appear to be quite hydrophobic. As all four of these fimbriae are immunologically distinct (23-25, 31, 36), the antigenic epitopes mediating serological specificity are probably outside the N-terminal regions of the subunits. In the case of some fimbrial subunit proteins, e.g. those of KS71A, F7, F7$_2$, Pap and F11 fimbriae, the conserved N-termini and C-termini have been shown to be more hydrophobic than the non-conserved regions of their primary structures (54). It has been suggested that the non-conserved regions are located at the outside of the subunits and also at the outside of the fimbriae (54). If this model structure holds for CFA/I, CS1, CS2 and CS4 fimbriae, it could account for the immunogenic diversity of these proteins. Indeed, the specific antigenic epitopes of these fimbriae may be related to quaternary structure (supramolecular structure) rather than to amino acid sequence.

The N-terminal amino acid homology of CS1, CS2, CS4 and CFA/I fimbrial subunits is indicative of a group of evolutionarily related fimbriae. According to Klemm (8) and de Graaf and Mooi (10), type 1 fimbriae, F7 to F12 fimbriae,

Table 3. Comparative N-terminal amino acid sequences of fimbrial subunits of human ETEC.

Fimbria	N-terminal amino acid sequence*						Reference
CFA/I**	V E K N I	T V T A S	V D P V I	D L L Q A	D G N A L	P S A V K	Klemm (45)
	V E K N I	T V T A S	V D P A I	D L L Q A	D G N A L	P S A V K	Hamers et al. (46)
CS1**	V E K T I	S V T A S	V D P T V	D L L Q S	D G S A L	P N D	Matthews (47)
CS2**	A E K N I	T V T A S	V D P T I	D L M Q S	D G T A L	P N D V N	Matthews (47)
**	A E K N I	T V T A S	V D P V I	D L L Q A			Klemm et al. (48)
CS4**	V E K N I	T V C A S	V D P T I	C I K Q A			Wolf et al. (36)
CS3***	A A G P T	L T K E L	A L N V L	S P A A L	D A T W A	P Q D N L	Boylan et al. (49)
							Hall et al.(cited in 49)
CS5***	A V T N G	Q L T F N	W Q G V V	P S A P V	T Q S S Q	P F V N G	Heuzenroeder et al. (50)

*Amino acid sequence homology between at least 3 out of 4 of the subunits for CFA/I, CS1, CS2 and CS4 fimbriae are highlighted in bold print.

**Amino acid sequencing data.

***Predicted amino acid sequence from nucleotide sequence of gene encoding structural protein

Pap fimbriae and K99 fimbriae form a distinct group based on the complete amino acid sequences of their structural subunits. Likewise, the K88 fimbrial series of antigenic variants forms a second distinct fimbrial group. The N-terminal amino acid sequence homology of the subunits of rod-shaped fimbriae of human ETEC (Table 3) suggests a third family comprising CFA/I, CS1, CS2 and CS4 fimbriae which have probably diverged from a common ancestor. In contrast, CS3 and CS5 fimbriae appear to belong to a distinct group or distinct groups, although the complete amino acid sequence of the CS5 fimbrial subunit is not known.

Amino acid analyses of purified fimbriae have revealed a lack of Cys residues in CS1, CS2, CS3 and CFA/I fimbrial subunits (Table 4). In contrast, CS4 fimbrial subunits have at least two Cys residues as shown in the first 20 N-terminal amino acids (Table 3). Thus, in the case of CFA/I, CS1, CS2 and CS3 fimbriae subunits must be held together by non--covalent bonds. The CFA/I, CS1, CS2 and CS3 fimbrial subunits also contain few aromatic amino acid residues. This contrasts with K88 fimbriae. CS1, CS2, CFA/I and type 1 fimbrial subunits each contain only two tyrosine residues, both of which are at the C-terminus in the case of the type 1 fimbrial subunit. It is, thus, of interest that aromatic amino acids have been implicated in the maintenance of the quaternary structure of some fimbriae (55,56).

GENETICS OF FIMBRIAE PRODUCTION IN HUMAN ETEC

General Aspects

Much work has been done in recent years on the genetic bases for the biosynthesis of fimbrial adhesins, although

Table 4. Amino acid compositions of CS1, CS2 and CFA/I subunit proteins.

Amino Acid	No. of amino acid residues per fimbrial subunit protein (Ref.)					
	CS1 (47)	CS2 (47)	CS2 (38)	CS2 (8)	CS2 (48)	CFA/I (45)
Asx	21	18	17	19	18	12
Thr	16	21	15	18	18	15
Ser	24	14	12	13	12	17
Glx	12	17	15	17	16	11
Pro	10	7	6	7	7	7
Gly	21	13	27	13	12	10
Ala	16	19	13	16	15	19
Cys	0	1	0	0	0	0
Val	17	10	10	12	12	19
Met	0	1	2	3	3	3
Ile	6	7	8	10	9	5
Leu	12	10	10	13	12	12
Tyr	2	2	2	3	3	4
Phe	3	4	3	4	4	2
His	3	1	1	2	2	1
Lys	7	9	7	10	9	8
Arg	2	3	3	4	4	1
Trp	0	0	ND*	ND*	0	1
Total	172	157	151	165	155	147

*Not determined

research on the genetics of production of fimbriae in human ETEC is perhaps not as advanced as in other systems, such as the K88 fimbriae of porcine ETEC, the P fimbriae of uropathogenic *E. coli* and type 1 fimbriae found on many enterobacteria.

The genes involved in the production of fimbriae on human ETEC generally reside on plasmids with some interesting exceptions. This has facilitated their analyses. Although there are differences in the organization of genetic determinants for fimbrial biogenesis, the data available for the genetic determinants of human ETEC fimbriae indicate features in common with the well studied K88, K99 and Pap fimbrial systems. For example, since fimbrial subunits must be transported across the outer membrane, a number of helper proteins possibly involved in translocation and assembly of fimbrial subunits have been identified and characterized in different fimbrial systems, e.g. K88 fimbriae (8,56) and Pap fimbriae (14,15,57,58). Accessory proteins of similar molecular weights to the gene products characterized for the K88 and Pap fimbrial operons have been identified among the gene products of the determinants for CFA/I, CS3 and CS5 fimbriae (50,51,59,60). Genes or gene products associated with expression of fimbriae on human ETEC strains are summarized (Table 5). Putative functions are assigned to these on the basis of knowledge of the K88, K99 and Pap fimbrial operons.

Table 5. Properties of human ETEC fimbrial polypeptides.

Fimbria	Gene	Polypeptide*	AA sequence**	Putative function of polypeptide	Reference
CFA/I:					
Region 1	cfaA	p23.5	+	Shuttle function in transport of subunits?	Hamers et al. (46)
	cfaB	p15	+	Structural subunit	Hamers et al. (46)
	cfaC	p85	+	Anchor in outer membrane?	Hamers et al. (46)
		p38.3	–	?	Willshaw et al. (51,61)
		p29.4	–	?	Willshaw et al. (51,61)
		p26.9	–	?	Willshaw et al. (51,61)
Region 2		p25.7	–	CFA/I Ras?=CfaD protein	Willshaw et al. (51,61)
		p14.9	–	?	Willshaw et al. (51,61)
		p13.8	–	?	Willshaw et al. (51,61)
		p30	+	CFA/I/=CfaD protein Rns cfaD	W. Gaastra (personal communication)
CFA/II:					
CS3		p97	–	Precursor of p94?	Manning et al. (59)
		p94		Anchor in outer membrane?	Manning et al. (59); Boylan et al. (60)
		p58	–	?	Manning et al. (59)
		p46.5	–	?	Manning et al. (59)
		p26/27	–	Precursor of p24?	Manning et al. (59); Boylan et al. (60)
		p24	–	Shuttle function in transport of subunits	Boylan et al. (60)
		p15	+	Structural subunit	Boylan et al. (59,60)
CS1/CS2	rns	p30	+	Rns	Caron et al. (62)
CFA/IV:					
CS5		p70	–	Anchor in outer membrane?	Heuzenroeder et al. (50)
		p46.5	–	?	Heuzenroeder et al. (50)
		p45	–	?	Heuzenroeder et al. (50)
		p31	–	?	Heuzenroeder et al. (50)
		p23	–	Structural subunit	Heuzenroeder et al. (50)
		p17	–	?	Heuzenroeder et al. (50)
		p14	–	?	Heuzenroeder et al. (50)

*Molecular mass in kilodaltons

**All amino acid sequences determined as translation products from the nucleotide sequences of cloned genes; + = sequence known, – = not determined.

CFA/II

The CFA/II fimbrial complex consists of three fimbrial antigens, viz. CS1, CS2 and CS3 (Tables 1 to 3). Strains expressing these fimbrial antigens may produce CS1 and CS3 fimbriae, CS2 and CS3 fimbriae, rarely CS2 fimbriae only, or CS3 fimbriae only (12). Strains bearing these fimbriae or fimbrial combinations show good adhesion to human small intestinal enterocytes (63-65). Expression of the CS fimbriae is plasmid-mediated, although the plasmids associated with production of these fimbriae in different strains appear to be quite heterogeneous with respect to molecular size, e.g.

44

22-115 MDa. The CS fimbriae phenotype of an ETEC strain is dependent upon the specific serotype and biotype of the host bacterium, which are useful phenotypic markers of predictive value for fimbrial expression (12). With the exception of one wild-type ETEC of serotype O139:H28 (66), only ETEC of serotype O6:K15:H16 or H- belonging to biotype A (67) produce CS1 fimbriae, whereas biotypes B, C and F of the same serotype produce CS2 fimbriae. Most strains of these biotypes also produce CS3 fimbriae. All other natural ETEC hosts of such CS fimbriae-associated plasmids produce only CS3 fimbriae irrespective of O serovar or O:K:H serotype and biotype (12).

Mobilization of CS-fimbriae-associated plasmids from a serotype O6:K15:H16 or H- background to other O6 serotypes or to other O servars or to *E. coli* K-12 gives rise to expression of only CS3 fimbriae in these new hosts, whilst the reverse mobilization of plasmids from a CS3-only wild-type background to serotype O6:K15:H16 or H- strains lacking a CS-fimbriae-associated plasmid gives rise to CS1 or CS2 fimbriae depending on the biotype (12). A K15 or H16 phenotype in combination with other O servars does not allow expression of CS1 or CS2 fimbriae in such *E. coli* hosts (68). These data taken together suggest that the expression of CS1 and CS2 fimbriae requires genetic information from the host chromosome and from the CS-fimbriae-associated plasmid (12).

CS3 fimbriae

Molecular cloning of the genetic determinant for expression of CS3 fimbriae has been achieved independently by three groups (59,60,69). The cloned DNA fragments determining CS3 fimbriae expression contain a cluster of genes encoding a number of proteins necessary for fimbrial biogenesis (59,60). Regions associated with expression of specific gene products have been defined by deletion mutagenesis or transposon mutagenesis, or both, coupled with analysis of plasmid--encoded proteins in minicells (Table 5). More recently the gene encoding the major subunit protein of CS3 fimbriae has been subcloned and sequenced (49). A possible promoter region and a putative ribosome-binding site were identified from the nucleotide sequence of the subcloned DNA fragment. The single open-reading frame encoded a polypeptide of 168 amino acids with a calculated molecular mass of 18.4 kDa. A potential signal peptidase cleavage site was identified between amino acid residues 21 and 22 (Ala-Ala). This yields a processed subunit of 16 kDa which is slightly higher than the apparent molecular weight determined by SDS polyacrylamide gel electrophoresis (Table 2). The first 27 amino acid residues of the mature CS3 protein identified by amino acid sequence analysis (Hall *et al.*, unpublished data cited in 49) agreed with those predicted from the nucleotide sequence of the subcloned DNA fragment.

As yet, neither the functions of the other gene products associated with CS3 fimbrial biogenesis nor the nucleotide sequences of these other genes and the predicted amino acid sequences of the translation products therefrom have been determined (Table 5). However, molecular mass similarities between gene products of the CS3 fimbrial determinant and those of other fimbrial determinants suggest roles analogous to those proposed for gene products of the K88, K99 and Pap

45

fimbrial gene clusters (10,15,56,57). These roles include being part of the structure anchoring the fimbriae to the outer membrane, or being involved in processing, polymerization, transport and assembly of the fimbrial subunits. Interestingly, both Manning *et al.* (59) and Boylan *et al.* (60) found that more DNA was required to encode the proteins identified by minicell analyses than there was available in their respective cloned CS3 fimbrial determinants, thereby implying overlapping genes. Such a phenomenon has not been described among the other fimbrial determinants which have been cloned and analyzed (10,56).

CS1/CS2 fimbriae

Molecular cloning of plasmid-located genetic determinants for the expression of CS1 and CS2 fimbriae has also been reported (69,70). In the case of the CS-antigen-associated plasmid NTP148, it was shown that the CS1/CS2 fimbrial determinant was located adjacent to the CS3 fimbrial determinant, but was not closely linked to genes for ST and LT production (69). This is in contrast to the organization of genes in a CFA/I fimbriae-associated plasmid in which DNA sequences encoding ST were found to be closely linked to one of the regions needed for CFA/I fimbriae production (see below; 51). The cloning and sub-cloning of CS1/CS2 fimbrial determinants demonstrated that the expression of CS1 and CS2 fimbriae determined by the cloned DNA fragments paralleled the host-related restriction of expression observed with the parental plasmids (pCS001 and NTP148) in *E. coli* K-12 and serotype O6:K15:H16 or H- ETEC strains of different biotypes (69-71).

Using a synthetic mixed 14-mer oligonucleotide probe corresponding to five of the N-terminal amino acids of the CS2 structural subunit protein determined by Klemm *et al.* (48), viz. E K N I T (Table 3), hybridization was observed with a chimaeric plasmid pCS200, comprising a 5.4 kbp *Hind*-III-generated, plasmid-derived DNA fragment cloned in vector plasmid pBR322, which mediated expression of CS1 and CS2 fimbriae in appropriate host strains (70). Thus, the data were consistent with an interpretation that the gene for the structural subunit of the CS2 fimbria resided on this plasmid. The nucleotide sequence corresponding to the N-terminal amino acid sequence of the CS1 fimbrial subunit (Table 3) would also be predicted to hybridize with the 14-mer DNA probe used by Boylan *et al.* (70). Taken together these observations provided at the time a possible explanation of the hybridization data for the 14-mer DNA probe to restriction endonuclease fragments encompassing more than one region of the cloned DNA sequences determining expression of CS1 and CS2 fimbriae.

However, subsequent studies on the pCS200 chimaera by Caron *et al.* (62) have shown that the cloned DNA sequences do not contain the gene for a CS2 structural fimbrial protein. They identified a CS-fimbriae-associated, plasmid-encoded gene that is required for expression of CS1 and CS2 fimbriae. The gene product termed Rns (Regulation of CS1 and 2) appears to be a regulatory protein that may act positively on the genes for production of these fimbriae encoded on the host chromosome. A chimaera called pEU2030, comprising a 1.2 kbp

46

*Hha*I fragment of plasmid pCS200 subcloned into vector plasmid
pUC18, contained only one open-reading frame of 795 bp. The
predicted translation product of the *rns* gene is a 265 amino
acid polypeptide with a molecular mass of 30 kDa and a pI of
10.1. The nucleotide sequence analysis of Caron *et al.* (62)
also provides an explanation of the DNA hybridization data of
Boylan *et al.* (70) in relation to their use of the mixed
14-mer oligonucleotide probe. A region identical in 13 of 14
bases to one of the components of the mixed oligonucleotide
probe was present 95 bp upstream of the ATG start codon of
the *rns* gene. Another homologous region, showing an 11 out of
14 bp match with a component of the mixed 14-mer probe, was
also found to lie outside the *rns* open-reading frame.

In appropriate plasmidless host backgrounds the pEU2030
chimaera allowed expression of CS1 or CS2 fimbrial antigens
with subunit molecular weights corresponding to those reported for native CS1 and CS2 fimbriae (Table 2). Moreover, the
host strain bearing the pEU2030 chimaera caused mannose--
resistant haemagglutination of bovine erythrocytes. These
data can only be interpreted as demonstrating that the genes
encoding the structural subunits of CS1 and CS2 fimbriae and
the genes for accessory proteins required for biogenesis must
be chromosomally located. This explanation is totally consistent with one model for CS1 and CS2 fimbriae expression
proposed by Smyth (12) to explain the host-restricted nature
of CS1 and CS2 fimbriae expression.

rns and the Rns protein

The CFA/II Rns protein shows significant amino acid
homology with the carboxy terminal halves of the amino acid
sequences of the AraC protein of *E. coli*, the regulatory
protein of the arabinose operon (62), of the RhaR and RhaS
proteins of the *E. coli* rhamnose operon (62), of the plasmid--
-encoded regulatory protein in *Yersinia enterocolitica,* VirF,
which controls the virulence regulon mediating calcium dependency, secretion of *Yersinia* outer membrane proteins (YOPs)
and pathogenicity (72) and, in particular, of the plasmid--
encoded regulatory protein in *Shigella flexneri*, VirF, which
controls expression of genes required for cell-to-cell spread
of the bacterium and bacterial invasion of epithelial cells
(Table 6) (73-75). Amino acid sequence homology between the
Rns and AraC proteins also includes the DNA-binding domain of
the *E. coli* AraC protein (amino acid residues 195 to 213, not
shown in Table 6) (62). The homology of the C-terminal amino
acid sequence of Rns with regions of the AraC protein that
appear to be significant in making functional DNA contacts
(62) and the fact that both the VirF proteins of *Y. enterocolitica* and *S. flexneri* have been shown to be transcriptional activators (72,75) suggests that the CFA/II Rns protein
acts directly to activate expression of CS1 and CS2 fimbriae.
Moreoever, the VirF protein of *Y. enterocolitica* has been
shown to possess DNA binding properties (G. Cornelis, this
Workshop).

Curiously the G + C content of the *rns* gene (28%) is low
for *E. coli* genes (50%). Moreover, codons used rarely in *E.
coli* occur frequently in the *rns* gene: AGA(G) for arginine,
AUA for isoleucine and GGA(G) for glycine (62). In genes

Table 6. Amino acid sequence homology between *Escherichia coli* Rns protein and AraC protein, *Yersinia enterocolitica* VirF protein and *Shigella flexneri* VirF protein*

```
                210                    220
E.c AraC    G I S V L S E R E D Q R I S Q A K L L L S T T R M P I
                210                    220
E.c. Rns        N F N Q I L M Q L R M S K A A L L L L E N S Y Q I
                      210                    220
S.f. VirF       T F Q Q I L L D I R M H H A A K L L L N S Q S Y I
                  210                220                    230
Y.e. VirF   G I S P R A W I S E R R I L Y A H Q L L L N G K M S I


                230                    240                    250
E.c. AraC   A T V G R N V G F D D Q L Y F S R V F K K C T G A S P
                230                    240                    250
E.c. Rns    S Q I S N M I G I S S A S Y F I R I F N K H Y G V T P
                  230                    240                    250
S.f. VirF   N D V S R L I G I S S P S Y F I R K F N E Y Y G I T P
                    240                    250
Y.e. VirF   V D I A M E A G F S S Q S Y F T Q S Y R R R F G C T P


                260                    270
E.c. AraC   S E F R A G C E E K V N D V A V K
                260
E.c. Rns    K Q F F T Y F K G G
                      260
S.f. VirF   K K F Y L Y H K K F
              260                    270
Y.e. VirF   S Q A R L T K I A T T G
```

*Homologies between 2, 3 or 4 sequences are highlighted in bold; amino acid residues are numbered from N-terminus of polypeptide; E.c., *Escherichia coli*; S.f., *Shigella flexneri*; Y.e., *Yersinia enterocolitica*.

which are not highly expressed, rare codons are frequently used. This is consistent with the proposed function of Rns as a regulatory protein. The G + C contents of other plasmid-located genes encoding virulence factors in ETEC are also low (ST1 = 31%; LT = 37%; CS3 = 36%) (62).

More than one *rns* gene may be present on plasmids mediating expression of CS1 and CS2 fimbriae. Boylan (76) independently cloned two *Hind*III fragments from plasmid pCS001 (71) and constructed two pBR322 chimaeras, one termed pCS200 from which Caron *et al.* (62) identified and sequenced *rns*, and pCS300 which mediated expression of CS2 fimbriae but not of CS1 fimbriae in appropriate wild-type backgrounds. The pCS001-derived DNA insert in the pCS300 chimaera has been extensively mapped for restriction endonuclease cleavage sites and differs in structure from the DNA insert in plasmid pCS200 (76; S. Smith, University of Dublin, unpublished data). Currently subcloning and sequencing is being undertaken to determine whether a *rns* gene exists in the pCS300 chimaera.

CFA/I

Transposon mutagenesis and restriction endonuclease cleavage site mapping of a plasmid, NTP113, involved in CFA/I fimbriae and ST production (77) revealed two widely separated regions designated regions 1 and 2 which are necessary for expression of CFA/I fimbriae (Table 5). The genetic determinants in these regions were cloned independently into compatible plasmid vectors (61). An *E. coli* strain carrying both chimaeras produced CFA/I fimbriae. In *E. coli* minicells, region 1 directed the expression of at least six polypeptides of which the CFA/I structural subunit and three other polypeptides appeared to be synthesized as precursors which underwent processing (51). Region 2 mediated expression of three polypeptides in minicells, one of which was essential for assembly of fimbriae (51).

Nucleotide sequencing of region 2 has revealed an open--reading frame encoding a protein of circa 30,000 daltons with a predicted amino acid sequence very similar to that of the CFA/II Rns protein of Caron *et al.* (62) controlling expression of CS1 and CS2 fimbriae (W. Gaastra, University of Utrecht, personal communication). Twenty-eight differences between the nucleotide sequences of the CS1/CS2 *rns* gene and the CFA/I *rns* gene resulted in 11 amino acid differences in the primary sequences of the respective *rns* gene products. The CFA/I *rns* gene when subcloned would also mediate expression of CS1 and CS2 fimbriae in appropriate host backgrounds, but not expression of CS4 fimbriae (H.R. Smith, PHLS Colindale, London, personal communication).

Thus, several antigenically distinct fimbriae on ETEC of human origin appear not only to comprise subunits with N--terminal amino acid sequence homology (Table 3), but some also share a similar Rns regulatory protein, indeed perhaps an evolutionarily derived common regulatory mechanism. In the case of CS1 and CS2 fimbriae the Rns protein is plasmid-encoded and regulates chromosomal genes, whereas with the CFA/I fimbrial system both the genes for production of CFA/I fimbriae and the *rns* gene are plasmid-located although in different regions. Moreover, there also appears to be a silent second copy of the *rns* gene in region 1 of the CFA/I plasmid (W. Gaastra, University of Utrecht, personal communication).

CFA/IV

Expression of the fimbrial antigens of the CFA/IV complex (CS4, CS5 and CS6) is plasmid mediated but a diversity of genetic linkages exists (Table 1). DNA sequences encoding the CS6 antigen of CFA/IV complex have been cloned from a plasmid (69). However, this has not yet led to the identification of the CS6 antigen morphologically on the surface of CFA/IV⁺ bacteria. In contrast, the genes for CS4 fimbriae expression have not as yet been successfully transferred to *E. coli* K-12 or cloned (36). The CS4 fimbrial determinant could be on a separate non-autotransferring plasmid from that bearing CS6 fimbriae-associated genes or be chromosomally located. Alternatively, as with the CS1/CS2 fimbrial system, regulatory functions necessary for CS4 fimbriae expression may be plasmid-located and may control expression of chromosomally located genes for fimbrial biogenesis in appropriate wild-type backgrounds.

The genes determining biosynthesis of CS5 fimbriae have been cloned (50). The CS5 fimbriae expressed from the cloned DNA sequences appeared to differ morphologically from the CS5 fimbriae described by Thomas *et al.* (31) in being wiry and of 3 nm diam. instead of rigid fimbriae of 6-7 nm diam. The wild-type strains examined by Heuzenroeder *et al.* (50) also had rigid rod- shaped fimbriae, possibly type 1 fimbriae, which did not react with antibodies directed against the 23,000 dalton CS5 fimbrial subunit protein.

With respect to this apparent anomaly in morphology it is of interest that Thomas *et al.* (31) noted the presence of curly wiry fimbriae on CS6[+] bacteria of some strains. Such curly fimbriae have also been described on ETEC of serotype 0148:H28 (34), a serovar known to produce CS6 antigen (Table 1). These wiry fimbriae look morphologically similar to fimbriae recently described by Olsen *et al.* (78), termed curli, as being fibronectin-binding fimbriae on *E. coli* strains associated with bovine mastitis.

Morphological differences in serologically identical fimbriae may be genetically determined. For example, differences in the morphology of F1C fimbriae and the degree of F1C fimbriation have recently been linked to the presence or absence of two accessory proteins, encoded by a cloned F1C genetic determinant, termed the FocH and FocI proteins (<u>f</u>imbriae of serotype <u>one</u> <u>C</u>) (79). These minor fimbrial proteins appear to serve some structural function. Deletion or mutation of the *focH* or *focI* locus resulted in changes in F1C fimbrial morphology varying from stubby rods to short or long rigid fimbriae. Moreover, deletion of a 4kbp HindIII region upstream of the *foc* gene cluster was associated with expression of wiry to curly F1C fimbriae. Thus it is possible that a mutation or deletion in the CS5 fimbrial determinant of the strain used by Heuzenroeder *et al.* (50) may account for the apparent discrepancy in the morphology of CS5 fimbriae compared with that described by Thomas *et al.* (31).

CODA

In the last few years great strides have been made in identifying and characterizing fimbriae present on strains of ETEC of human origin. As with strains of animal origin these studies have revealed that several antigenically different types of fimbriae can be expressed on human ETEC. Much effort is being devoted to the cloning of genes mediating fimbrial biogenesis. Novel control mechanisms for expression of fimbriae on ETEC of human origin have been revealed. Indeed the biogenesis of fimbriae in such strains appears to be more complex than revealed with the K88, K99 and Pap fimbrial operons. Much still requires to be done to reveal and understand the precise nature of the control mechanisms the numbers of accessory proteins involved in the biogenesis of fimbriae on ETEC of human origin.

REFERENCES

1.   Sack, R.B. 1975. Ann. Rev. Microbiol. 29:333-363.
2.   Merson, M.H., R.E. Black, M.U. Khan and I. Huq. 1980. In Cholera and Related Diseases (Ouchterlony, D. and J. Holmgren, eds.) 43rd Nobel Symposium, pp. 34-35, S. Karger, Basel.
3.   Black, R.E., M.H. Merson, I. Huq, A.R. Alim and M. Yunus. 1981. Lancet i, 141-143.
4.   Abe, H., S. Ichiki, S. Hashimoto, H. Nalkano, K. Sato, T. Kanda, Y. Yanai, T. Tsukamoto, Y. Kinoshita, M. Arita, T. Honda, Y. Takeda and T. Miwatani. 1984. J. Diar. Dis. Res. 2:83-87.
5.   Levine, M.M. 1984. In Bacterial Vaccines. (Germanier, R. ed.), pp. 187-235, Academic Press, London.
6.   Holmgren, J. 1985. in The Virulence of *Escherichia coli:* Reviews and Methods (Sussman, M. ed.) Special Publications of the Society for General Microbiology, 13:177-191, Academic Press, London.
7.   Jones, G.W. and R.E. Isaacson. 1984. C.R.C. Crit. Rev. Microbiol. 10:229-260.
8.   Klemm, P. 1986. Rev. Infect. Dis. 7:321-340.
9.   Parry, S.H. and D.M. Rooke. 1985. In The Virulence of *Escherichia coli:* Reviews and Methods (Sussman, M. ed.) Special Publications of the Society for General Microbiology, 13:79-155, Academic Press, London.
10.   de Graaf, F.K> and F/R/ Mooi. 1986. Adv. Microb. Physiol. 28:65-143.
11.   Ofek, I.O. and E.H. Beachey. In Microbial Interactions (Reissig, J.L., ed.) Receptors and Recognition Series B. pp. 1-29. Chapman and Hall, London.
12.   Smyth, C.J. 1986. In Antigenic Variation in Infectious Diseases (Birkbeck, T.H. and Penn, C.W., eds) Special Publications of the Society for General Microbiology, 19:95-125, IRL Press, Oxsford.
13.   Mirelman, D. and I. Ofek. 1988. In Microbial Lectins and Agglutinins: Properties and Biological Activity (Mirelman, D. ed.) pp. 1-19, John Wiley and Sons, New York.
14.   Lark, D.L., S. Normark, B.-E. Uhlin and H. Wolf-Watz. 1986. Protein-carbohydrate interactions in Biological Systems. FEMS Symposium No. 31, Academic Press, London.
15.   Normark, S., M. Baga, M. Goransson, F.P. Lindberg, B. Lund, M. Norgren and B.-E. Uhlin. 1986. In: Microbial Lectins and Agglutinins: Properties and Biological Activity (Mirelman, D. ed.) pp. 113-143, John Wiley and Sons, New York.
16.   Abraham, S.N., J.D. Goguen, D. Sun, P. Klemm and E.H. Beachey. 1987. J. Bacteriol. 169:5530-5536.
17.   Oudega, B., F. de Graaf, L. de Boer, D. Bakker, C.E.M. Vader, F.R. Mooi and F.K. de Graaf. 1989. Mol. Microbiol. 3:645-652.
18.   Lindberg, F., B. Lund and S. Normark. 1986. Proc. Natl. Acad. Sci. USA. 83:1891-1895.
19.   Lindberg, F., B. Lund, L. Johansson and S. Normark. 1987. Nature, London 328:84-87.
20.   Krogfelt, K.A. and P. Klemm. 1988. Microb. Pathog, 4:231-238.
21.   Evans, D.G., R.P. Silver, D.J. Evans, Jr., D.G. Chase and S.L. Gorbach. 1985. Infect. Immun. 12:656-667.
22.   McConnel, M.M., H. Chart, A.M. Field, M. Hibberd and B.

Rowe. 1989. J. Gen. Microbiol. 135:1135-1144.

23. Evans, D.G. and D.J. Evans, Jr. 1978. Infect. Immun. 21:638-647.

24. Cravioto, A., S.M. Scotland and B. Rowe. 1982. Infect. Immun. 36:189-197.

25. Smyth, C.J. 1982. J. Gen. Microbiol. 128:2081-2096.

26. Smyth, C.J. 1984. FEMS Microbiol. Lett. 21:51-57.

27. Honda, T., M.M.A. Khan, Y. Takeda and T. Miwatani. 1983. FEMS Microbiol. Lett. 17:273-276.

28. Honda, T., M. Arita and T. Miwatani. 1984. Infect. Immun. 43:959-965.

29. McConnel, M.M. and B. Rowe. 1989. J. infect. Dis. 159:582-585.

30. Thoman, L.V. and B. Rowe. 1982. Med. Microbiol. Immunol. 171:85-90.

31. Thomas, L.V. , M.M. McConnel, B. Rowe and A.M. Field. 1985. J. Gen. Microbiol. 131:2319-2326.

32. Tacket, C.O., D.R. Maneval and M.M. Levine. 1987. Infect. Immun. 55:1063-1069.

33. McConnell, M.M., L.V. Thoman, S.M. Scotland and B. Rowe. 1986. Curr. microbiol. 14:51-54.

34. Knutton, S., D.R. Lloyd and A.S. McNeish. 1987. Infect. Immun. 55:86-92.

35. Levine, M.M. 1987. J. Infect. Dis. 155:377-389.

36. Wolf, M.K., G.P. Andrews, B.D. Tall, M.M. McConnel, M.M. Levind and E.C. Boedeker. 1989. Infect. Immun. 57:164-173.

37. Evans, D.J., Jr., D.G. Evans, S.R> Diaz and D.Y. Graham. 1989. J. Clin. Microbiol. 26:1626-1629.

38. Sjoberg, P.-O., M. Lindahl, J. Porath and T. Wadstrom. 1988. Biochem. J. 255:105-111.

39. Mouricout, M. and R. Julien. 1986. FEMS Microbiol. Lett. 37:145-149.

40. Wadstrom, T. and T.J. Trust. 1984. in Medical Microbiology (Easmon, C.S.F. and J. Jeljaszewicz, eds.). Vol. 4, 283-334, Acad. Press London.

41. Guinee, P.A.M. and W.H. Jansen. 1979. Infect. Immun. 23:700-705.

42. Soderlind, O. E. Olsson, C.J. Smyth and R. Mollby. 1982. Infect. Immun. 36:900-906.

43. Brinton, C.C>Jr. 1965. Trans. N.Y. Acad. Sci. 27:1003-1054.

44. Steven, A.C., M.E. Bisher, B.L. Trus, D. Thomas, J.M. Zhang and J.L. Cowell. 1986. J. Bacteriol. 167:968-974.

45. Klemm, P. 1982. Eur. J. Biochem. 124:339-348.

46. Hamerz, A.M., H.L. Pel, G.A. Willshaw, J.G. Kusters, B.A.M. van der Zeijst and W. Gaastra. 1989. Microb. Pathog. 6:297-309.

47. Matthews, H.M.C. 1989. MSc Thesis, Trinity College, University of Dublin.

48. Klemm, P., W. Gaastra, M.M. McConnell and H.R. Smith. 1985. FEMS Microbiol. Lett. 26:207-210.

49. Boylan, M., C.J. Smyth and J.R. Scott. 1988. Infect. Immun. 56:3297-3300.

50. Heuzenroeder, M.W., B.L. Neal, C.J. Thoman, R. Halter and P.A. Manning. 1989. Mol. Microbiol. 3:303-310.

51. Willshaw, G.A., H.R. Smith, M.M. McConnel and B. Rowe. 1985. Plasmid 13:8-16.

52. Levine, M.M., P. Ristaino, G. Marley, C. Smyth, S. Knutton, E. Boedeker, R. Black, C. Young, M.L. Clements,

C. Cheney and R. Patnaik. 1984. Infect. Immun. 44:409-420.

53. McConnel, M.M., L.V. Thoman, G.A. Willshaw, H.R. Smith and B. Rowe. 1988. Infect. Immun. 56:1974-1980.

54. van Die,I., W. Hoekstra and H. Bergmans. 1987. Microb. Pathog. 3:149-154.

55. Watts, T.H., C.M. Kay and W. Paranchych. 1983. Biochemistry 22:3640-3646.

56. Mooi, F.R. and F.K. de Graaf. 1985. Curr. Top. Microbiol. Immun. 118:119-138.

57. Smyth, C.J. 1988. in Immunochemical and Molecular Genetic Analysis of Bacterial Pathogens (Owen, P. and T.J. Foster, eds), 13-25, Elsevier, Amsterdam.

58. Hultgren, S.J., F. Lindberg, G. Magnusson, J.M. Tennent and S. Normark. 1989. in Molecular Mechanisms of Microbial Adhesion (Switalski, L., M. Hook and E. Beachey, eds), 36-43, Springer-Verlag, New York.

59. Manning, P.A., K.N. Timmis and G. Stevenson. 1985. Mol. Gen. Genet. 200:322-327.

60. Boylan, M., D.C. Coleman and C.J. Smyth. 1987. Microb. Pathog. 2:195-209.

61. Willshaw, G.A., H.R. Smith and B. Rowe. 1983. FEMS Microbiol. Lett. 16:101-106.

62. Caron, J., L.M. Coffield and J.R. Scott. 1989. Proc. Natl. Acad. Sci. USA 86:963-967.

63. Knutton, S., D.R. Lloyd, D.C.A. Candy and A.S. McNeish. 1984. Infect, Immun. 44:514-518.

64. Knutton, S., D.R. Lloyd, D.C.A. Candy and A.S. McNeish. 1984. Infect, Immun. 44:519-527.

65. Knutton, S., D.R. Lloyd, D.C.A. Candy and A.S. McNeish. 1984. Infect, Immun. 48:824-831.

66. Scotland, S.M., M.M. McConnel, G.A. Willshaw, B. Rowe and A.M. Field. 1985. J. Gen. Microbiol. 131:2327-2333.

67. Scotland, S.M., R.J. Gross and B. Rowe. 1977. J. Hyg. Camb. 79:395-403.

68. Twohig, J., M. Boylan and C.J. Smyth. 1988. FEMS Microbiol. Lett. 56:327-330.

69. Willshaw, G.A., H.R. Smith, M.M. McConnel and B. Rowe. 1988. FEMS Microbiol. Lettl 49:473-478.

70. Boylan, M., D.C. Coleman, J.R> Scott and C.J. Smyth. 1988. J. Gen. Microbiol. 134:2189-2199.

71. Boylan, M. and C.J. Smyth. 1985. FEMS Miccrobiol. Lett. 29:83-89.

72. Cornelis, G., C. Sluiters C. Lambert de Rouvroit and T. Michelis. 1989. J. Bacteriol. 171:254-262.

73. Sakai, T. C. Sasakawa, S. Makino, K. Kamata and M. Yoshikawa. 1986. Infect. Immun. 51:476-482.

74. Sakai, T., C, Sasakawa and M. Moshidawa. 1988. Mol. Microbiol. 2:589-597.

76. Boylan, M.H. 1987. PhD Thesis, Trinity College, University of Dublin.

77. Smith, H.R., Willshaw, G,A. and B. Rowe. 1982. J. Bacteriol. 149:264-275.

78. Olsen, A., A. Jonsson and S. Normark. 1989. Nature, London 338:652-655.

79. van Die, I., N. Riegman, J. van Oosterhout, H. Bergmans, A. van der Zee, H. Bes and W. Hoekstra. 1989. Microbial Surface Components and Toxins in Relation to Pathogenesis, Plenum Press.

REGULATION OF THE BIOSYNTHESIS OF K99 AND 987P FIMBRIAE

Frits K. de Graaf, Marjan W. van der Woude and
Pia Klaasen

Department of Molecular Microbiology, Faculty of
Biology, Vrije Universiteit, de Boelelaan 1087
1081 HV Amsterdam, The Netherlands

INTRODUCTION

The fimbrial adhesins K99 and 987P are produced by enterotoxigenic *Escherichia coli* strains that provoke neonatal diarrhea in calves and piglets. The biosynthesis of these fimbriae is affected by environmental conditions. Optimal production is achieved at 37°C and high specific growth rate (μ). At lower temperatures or lower specific growth rates the production of fimbriae per unit of biomass rapidly decreases (9). In this study a combination of physiological and genetical approaches was used to elucidate the mechanisms of regulation of fimbriae expression.

REGULATION OF K99 EXPRESSION

The genetic determinant for the biosynthesis of K99 fimbriae has been cloned into pBR322 (3). Experiments with the recombinant plasmid, pFK99, in *E. coli* K-12 have shown that the effect of the specific growth rate, demonstrated with wild-type strains, also occur in cells harbouring pFK99. This indicates that regulatory sequences that might be involved in the response of the cells to changes in the specific growth rate, with respect to K99 production, are probably contained within pFK99. The genetic map of the cloned K99 determinant is shown in Fig. 1. The K99 operon encodes the K99 fimbrial subunit (FanC), the proteins involved in the translocation of fimbrial subunits across the cell envelope (FanD and E), as well as the minor fimbrial subunits (Fan-F-H), involved in the biosynthesis of the fimbriae. Upstream of fanC two open reading frames were detected, fanA and fanB, encoding the trans-acting regulatory proteins FanA and FanB (5). Indications were obtained that these regulators, most likely in concert, function as anti-terminators of transcription in the K99 regulatory region (6). A detailed analysis of the transcription signals in the regulatory region encompassing fanA and fanB revealed that a strong promoter P_A, and a moderate promoter P_B are located upstream of fanA and fanB, respectively, while no promoter activity was detected in the intercistronic region between fanB and fanC (6). Furthermore, terminators of transcription which are affected by FanA

Microbial Surface Components and Toxins in Relation to Pathogenesis
Edited by E.Z. Ron and S. Rottem, Plenum Press, New York, 1991

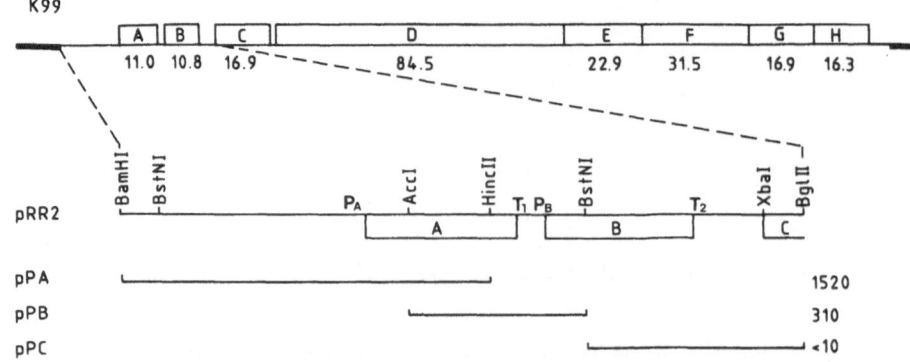

Fig. 1.   A. Genetic map of the K99 operon (pFK99), B. The transcriptional organization of the K99 regulatory region (pRR2), C. pRR2-derived clones used for measuring promoter activity (pPA, pPB and pPC).

and FanB are located in the intercistronic region between fanA and fanB (T_1) and between fanB and fanC (T_2). A third and strong terminator (T_3) is present between fanC and fanD.

To test whether the transcriptional activity of the regulatory region controlling the expression of the fimbrial subunit FanC is affected by temperature and/or the specific growth rate, a DNA segment encompassing fanA, fanB and the 5' end of fanC was cloned into the transcription vector pKO-1. This vector contains the galactokinase (GalK) gene as assayable marker but does not contain a promoter upstream of the galK gene (4). Therefore, expression of GalK activity is dependent on the transcriptional activity of DNA segments inserted in unique restriction sites preceding the galK gene. The constructed recombinant plasmid was designated pRR2 (Fig. 1). Plasmid pRR2 was used to transform a wild-type K99-producing strain. Subsequently, the galactokinase activity and the K99 production were measured at different specific growth rates (Fig. 2). The growth rate-dependent expression of K99 fimbriae appeared to be similar to that in wild-type K99 producing cells without pRR2. Apparently, an excess of regulatory proteins does not affect this phenomenon.

The galactokinase activity expressed by pRR2 at different growth rates was corrected for the plasmid copy number and the background GalK activity, due to the expression of the chromosomal galK gene (<10 U.) Rather low GalK activity was measured at μ's below 0.2 h^{-1}, where virtually no fimbriae are produced. An intermediate level of GalK activity was measured at μ's between 0.2 and 0.5 h^{-1} concomitant with a slight increase in K99 production. A high level of GalK activity was observed at $\mu > 0.5$ h^{-1} where K99 production strongly increases (Fig. 2). The experiments show that transcription of the regulatory region is growth-rate dependent.

To test whether this phenomenon could be attributed to a growth rate-dependent activity of the promoter P_A and/or P_B, small DNA fragments containing these promoters (Fig. 1) were cloned into pKO-1 and the galactokinase activity of strains harbouring one of these plasmids was measured at different growth rates (Fig. 3). The activity of P_A increases with

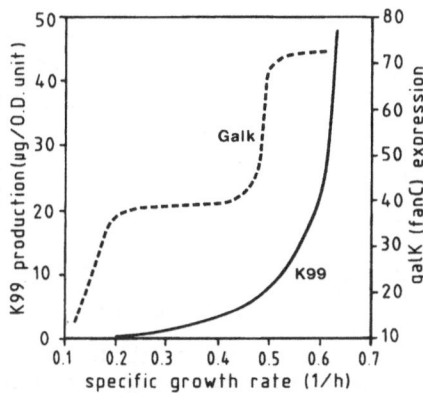

Fig. 2.   Growth rate-dependent synthesis of K99 fimbriae and
           galactokinase by wild-type K99-producing strains
           harbouring pRR2.

increasing growth-rate while the P_B activity remains constant. Although these experiments show that promoter P_A plays
an important role in the growth rate-dependent synthesis of
K99 fimbriae, the differences in the expression of the wild
type K99 plasmid, pRR2 and pPA, respectively, in relation to
the specific growth rate of the bacteria, probably indicate
that the growth rate-dependent response with respect to K99
production is a more complex phenomenon and may include
additional effects on the anti-terminator activity of FanA
and FanB or on mRNA stability (post- transcriptional regulation). In addition, changes in the cell envelope structure
may effect the assembly of fimbriae.

     The effect of temperature on the biosynthesis of K99
fimbriae can be explained in part by the growth rate -depend-
ent expression of these fimbriae. At lower incubation temperatures the specific growth rate decreases, which results in a
lower rate of K99 synthesis. Apart from this, temperature
itself has a clear effect on the biosynthesis of K99 fimbriae. Fig. 4 shows the results of an experiment in which a
wild-type K99-producing strain was grown in the chemostat at
a fixed specific growth rate of 0.4 h^{-1} at various temperatures. At temperatures below 32°C very little synthesis of
K99 fimbriae occurs. Above 32°C the synthesis gradually

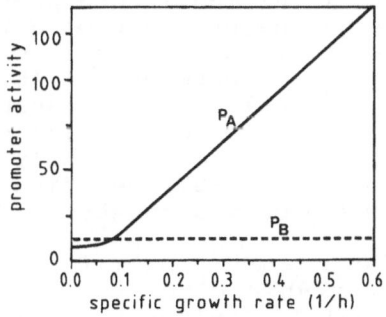

Fig. 3.   Growth rate-dependent activity of the P_A and PB
           promoters.

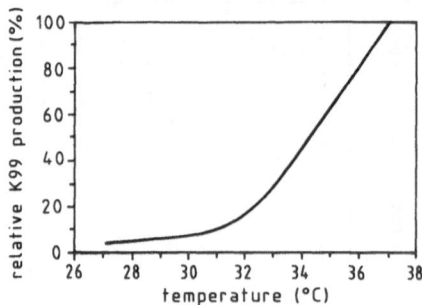

Fig. 4.   Temperature-dependent synthesis of K99 fimbriae by
           wild-type strains growing in the chemostat at a
           fixed specific growth rate of 0.4 h^{-1}.

increases and reaches an optimum at 37°C. When this experiment was repeated with the wild-type strain containing both
the K99 plasmid as well as multicopies of pRR2, the effect of
temperature disappears and expression of K99 fimbriae remains
at a constant level. The most likely explanation for this
result is that the concentration of some host factor involved
in temperature-control of production and capable of binding
to regulatory sequences encoded by pRR2 is too low to affect
K99 expression in the presence of multi-copies of pRR2.

REGULATION OF 987P EXPRESSION

     DNA encoding   the   biosynthesis of   987P fimbriae was
cloned using the cosmid vector pHC79, followed by subcloning
of the 987P determinant into pBR322 (2; Fig. 5). Prolonged
cultivation of cells harbouring the recombinant plasmid
pPK180 frequently results in the loss of the ability of the
cells to synthesize 987P fimbriae. This instability is, in
part, due to the appearance of plasmidless cells, but in most
cases deletions within pPK180 occurred. Analysis of some
deletion derivatives indicated that in general a DNA fragment
of similar size was deleted in all mutants. One of the deletion mutants was used to further characterize the deletion by
nucleotide sequence analysis. It appeared that on one side
the deleted DNA was flanked by an IS 1 element while on the
other side the deletion extended into the promoter region
upstream of the 987P subunit gene (fapC, Fig. 5). Subsequent
analysis of the DNA region containing the IS 1 element showed
that in fact two IS 1 elements were present in an inverted
orientation indicating the presence of a transposon. Furthermore, it was shown that this transposon carries the gene
encoding the heat-stable enterotoxin (ST$_A$). Probably, this
ST$_A$ transposon is identical to the transposon Tn1681, previously found in enterotoxigenic E. coli strains (8). Because
IS 1-induced deletions were most likely to be the cause of
the frequent appearance of 987P-negative cells, a 987P clone
without the ST$_A$ transposon was constructed.

     To that purpose a BamHI - BalI fragment was deleted from
pPK180. The resultant 987P clone contains only a small part
of one of the original IS 1 elements (Fig. 5). Surprisingly,

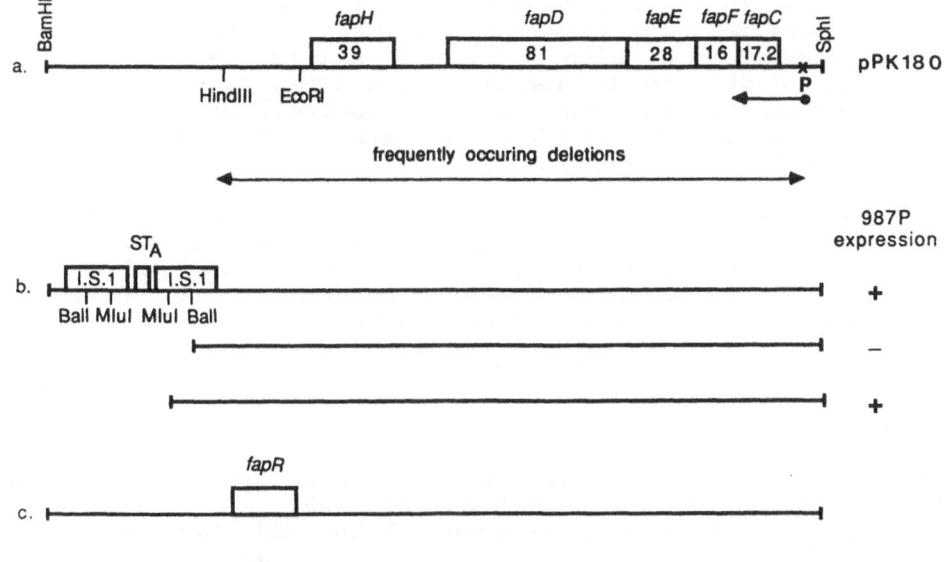

Fig. 5.   A. Genetic map of the 987P operon contained in
pPK180. The molecular mass of the various gene
products is indicated in kiloDalton. The arrow
indicates the position of the promoter and the
direction of transcription of the fimbrial subunit
gene (fapC). The size of frequently occurring deletions in pPK180 is indicated. B. The position of
t....ST$_A$ transposon present on pPK180, two pPK180--
derived clones containing a part of the right-hand
IS 1 element, and expression of 987P fimbriae. C.
The position and the direction of transcription of
the regulatory gene fapR contained in the 987P
operon.

this clone was not capable to produce 987P fimbriae indicating that sequences contained within the ST$_A$ transposon are
essential for 987P expression. Construction of a slightly
larger clone by removal of a BamHI-MluI fragment from pPK180
resulted in a 987P-positive clone. These results suggest that
at least the part of the IS 1 DNA which is located between
MluI and BalI, plays an important role in the expression of
the 987P gene cluster.

Previous analysis of the 987P gene cluster in *E. coli*
minicells, did not indicate whether any structural gene(s)
are present on the HindIII-EcoRI fragment positioned adjacent
to the ST$_A$ transposon in pPK180. On the other hand Tn5 insertions in this DNA segment result in a strong reduction of
987P expression. It has been suggested that this region could
be involved in the regulation of 987P expression (2). Sequence analysis of 987P DNA flanking the right-hand IS 1
element revealed the presence of an open reading frame,
designated fapR. This gene is transcribed from the left to
the right (Fig. 5), opposite to the direction of transcrip-

tion of the 987P subunit gene (fapC). The primary structure of FapR was deduced from the nucleotide sequence, and was found to possess a high degree of similarity with proteins recently identified as positive regulators of transcription, in particular the rns gene product controlling expression of CFA II fimbriae and the VirF protein controlling virulence markers in *Shigella dysentheria* (1,7).

The fapR gene transcript possesses a relatively strong ribosome binding site. However, a consensus sequence for an *E. coli* promoter was not observed. It is supposed that DNA sequences within the IS 1 element may function as enhancer of transcription of the fapR gene and that FapR functions as a positive regulator of 987P expression.

REFERENCES

1. Caron, J., L.M. Coffield, L.M. and J.R. Scott, J.R. 1989. Proc. Natl. Acad. Sci. USA 86: 963-967.
2. De Graaf, F.K. and P. Klaasen . 1986. Mol. Gen. Genet: 204: 75-81.
3. De Graaf, F.K., B.E. Krenn, and P. Klaasen, P. 1984. Infect. Immun. 43: 508-514.
4. McKenney, K., H. Shimatake, D. Court, U. Schmeissner, C. Brady and M. Rosenberg, M. 1981. In: Chirikjian, J.C., and Papas, T.S. (eds... Gene amplification and analysis, vol. II. Elsevier, North Holland, pp. 383-415.
5. Roosendaal, E., M. Boots, M. and F.K. de Graaf, F.K. 1987. Nucleic Acids Res. 15: 5973-5984.
6. Roosendaal, B., J. Damoiseaux, W. Jordi and F.K. de Graaf, 1989. Mol. Gen. Genet. 215: 250-256.
7. Sakai, T., C. Sasakawa and M. Yoshikawa 1988. Molec. Microbiol. 2: 589-597.
8. So, M., F. Heffron and B.J. McCarthy. 1979. Nature 277: 453-456.
9. Van Verseveld, H.W., P. Bakker, T. van der Woude, C. Terleth and F.K. de Graaf. 1985. Infect. Immun. 49: 159-163.

ADHERENCE PILI OF *ESCHERICHIA COLI* O78

DETERMINE HOST AND TISSUE SPECIFICITY

Eliora Z. Ron, Zohar Yerushalmi, and Michael W. Naveh

Department of Microbiology
Tel-Aviv University
Tel-Aviv, 69978, Israel

INTRODUCTION

Virulent *Escherichia coli* strains are involved with a variety of clinical diseases and in a broad spectrum of hosts. Often, the types of virulence factors a strain produces determine its ability to survive and multiply in certain organs, or in certain hosts. For example, the presence of a polysaccharide capsule enables the virulent strain that produces it to multiply in the blood stream and vital organs.

We are interested in the factors that determine the capability of virulent strains to multiply in specific hosts or specific tissues. For this purpose we have concentrated on one pathogenic *E. coli* strain - serotype O78 - that is special in its ability to cause a large variety of clinical syndromes and in a broad spectrum of hosts. We have been studying the genetics of the virulence factors in this strain and compared strains that are virulent to mammals and to poultry. One factor that was intensively studied is the adherence pili, as they determine the site of host recognition. The results presented in this communication indicate that there are differences between avian strains and strains isolated from humans, and these differences are important in determining host specificity.

PROPERTIES OF *E. COLI* O78 STRAINS

Virulent strains of *Escherichia coli* are the causative agents of intestinal infections as well as infections of the urinary tract, meningitis and sepsis. The intestinal infections are localized diseases, and there are certain serotypes (O18, O111 and others) that are commonly associated with them. As a rule, these strains are not encapsulated, produce enterotoxin and one or more types of adherence pili. Other serotypes - (such as O2, O4 and O6) have been frequently isolated from urinary tract infections, meningitis and systemic diseases - bacteremia and sepsis. These strains usually produce polysaccharide capsules (K1, K5 etc.) and many of

Microbial Surface Components and Toxins in Relation to Pathogenesis
Edited by E.Z. Ron and S. Rottem, Plenum Press, New York, 1991

61

them express pili of the P-type that recognize gal-gal disaccharides.

Strains of *E. coli* O78 serotype are special, as they are not restricted to enteric or systemic diseases, but are the causative agents of a broad spectrum of diseases, and in a variety of hosts. Thus, strains of *E. coli* O78 are frequently associated with diarrhea in humans and calves, and on the other hand, they are frequently found in meningitis and sepsis of babies, lambs and chicks. The broad host specificity of O78 strains, in addition to the diverse kinds of diseases that they cause, make it a good model system for studying the role of the various virulence factors in pathogenicity.

The structure of the O78 antigen has been determined (1). There is no detectable polysaccharide capsule (F. and I. Orskov, personal communication) even in the septicemic strains. However, there appears to be an "LPS-like" capsule, as in the case of O111 (2) (Grossman, Tchelet and Ron, unpublished), in which the basic structure of the "capsular" material is identical to that of the O78 antigen (1).

Enterotoxigenic O78 strains can be isolated from cases of diarrhea in humans and in animals. The O78 strains isolated from cases of human diarrhea have been shown to produce CFA/I pili as well as LT and ST enterotoxins (3). Other enterotoxigenic O78 strains were isolated from diarrhea in calves, and were shown to carry K99 or "K88-like" CS31A pili (4).

Septicemic O78 strains have been isolated from human cases of neonatal meningitis and from bacteremia in calves. The O78 strains from human neonatal meningitis carry *colV* plasmids (5,6) and produce P-pili of 20 kd (T. Korhonen, personal communication; Yerushalmi and Ron, unpublished). Our isolates from a septicemic disease of lambs also carry *colV* plasmids, but adhere to the host with K99 pili (Yerushalmi, Ron and Naveh, unpublished).

E. coli O78 is also a very important pathogen of poultry - chicks and turkeys, causing a septicemic disease with high morbidity and mortality. This disease is apparently initiated by bacterial invasion via the trachea followed by spreading to the air-sac and vital organs. Chicks are very sensitive to the virulent *E. coli* strains, in contrast to mice that are not easily infected (LD50 being about 10^7 or higher) and are, therefore useful as a model system to analyze the role of the various virulence factors in determining host and tissue specificity.

HOST SPECIFICITY OF AVIAN STRAINS AND STRAINS FROM HUMAN ORIGIN

In order to examine the virulence factors of the various O78 isolates, we collected a large number of pathogenic strains and compared their virulence by injecting i.p. into 5-day-old chicks. The results are presented in Fig. 1, and indicate that there is a good correlation between the degree of virulence and the original host and disease. Thus, non-

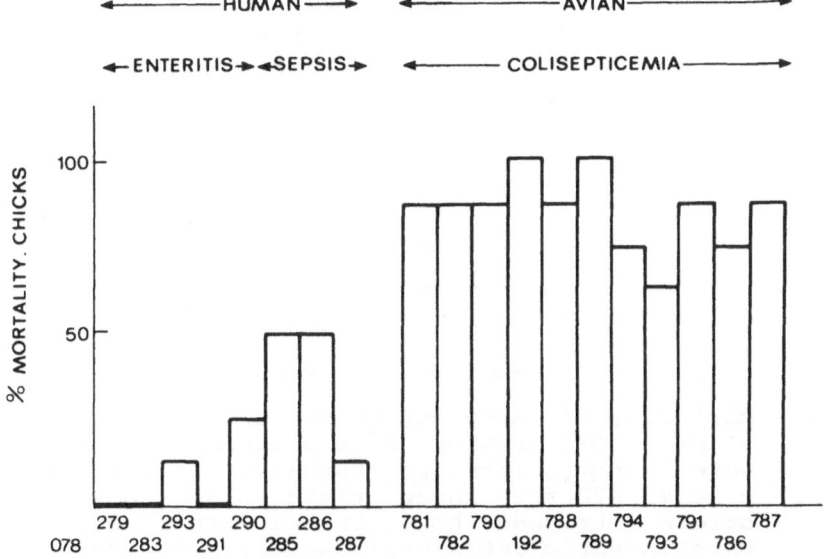

Fig. 1.   Mortality of chicks infected with *E coli* O78
           strains. One-day-old chicks were injected intra-
           peritoneally with 10^4 bacteria of various isolates
           of *E. coli* O78. The source of each isolate is indi-
           cated at the top of the figure and its serial num-
           ber at the bottom) The results presented summarize
           mortality 72 hours after inoculation.

toxigenic strains from healthy people, or enterotoxigenic
strains isolated from human diarrhea, were avirulent to
chicks. Strains carrying *colV* plasmids, isolated from human
neonatal meningitis were virulent, but not as virulent as
strains isolated from avian septicemia. The high virulence of
avian strains in infecting chicks could indicate that these
strains are more virulent than other *E. coli* strains, or that
they are better adapted to the avian host. These two possibilities were further checked by comparing an avian strain with
a similar human strain in an avian host and a mammalian host.
The two bacterial strains were of serotype O78 and carried
colV plasmid. Upon i.p. injection to chicks the avian strain
was more virulent than the strain of human origin but there
was no difference in virulence when the two strains were
injected into mice (Fig. 2). These results further supports
the finding that there exists a significant host-specificity
in *E. coli* O78 strains.

PILI FROM AVIAN *E. COLI* O78 STRAINS

     One factor that could play an important role in deter-
mining host specificity is adherence pili, as they are the
site of recognition. All the virulent O78 strains from
mammals were found to express CFA/I, P or K99 adherence pili.
Many of the avian strains were also piliated (7,8), and
several produce type I pili (9). We have examined several
piliated strains of avian *E. coli* O78. Their pili did not
hemagglutinate any of the erythrocytes that we examined, but

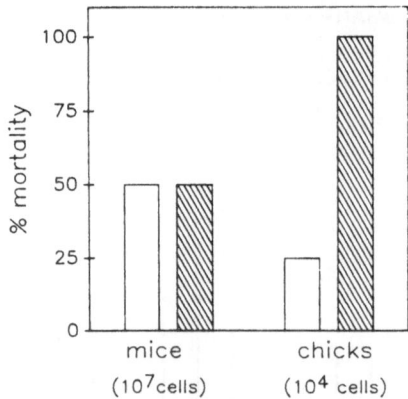

Fig. 2.   Mortality of mice and chicks after infection with
           E. coli O78 strains. Mice weighing about 30 grams
           were injected intra-peritoneally with 10^7 cells of
           E. coli O78 from human origin (blank box) or avian
           origin (stripped box). Chicks were 5 days old and
           were injected intra-peritoneally with 10^4 bacteria.
           Mortality was recorded 4 days following inocula-
           tion.

there were indications that they are important for pathogenesis (10). These pili were named by us "AC/I" (avian coli
#I) pili and have an apparent molecular weight of 18 kd (11).
It should be noted that Dr. van den Bosch has obtained evidence showing that avian *E. coli* strains produce F11 pili,
that are also 18 kd in molecular weight. The pili that we see
do not hemagglutinate human RBC or bind gal-gal disaccharides, and in this respect they are different from F11 pili.
They could, however, be of a similar type, and further studies are now in progress to determine the degree of relatedness of AC/I pili and F11 pili.

HOST AND TISSUE SPECIFICITY OF AC/I PILI

     Several experiments indicated that AC/I pili are impor-
tant for adherence to avian epithelial cells, both *in vitro*
and *in vivo* (10,11). An example of one such experiment is
presented in Fig. 3.

     In order to find out whether the adherence mediated by
AC/I pili is specific for avian tissues, we compared O78 bacteria carrying AC/I pili to bacteria carrying CFA/I pili in
adherence to avian tracheal cells and human buccal cells. The
results shown in Table 1. indicate that AC/I pili adhere
preferentially to the avian cells, in contrast to CFA/I pili.

     The next question was whether there is also a "tissue
specificity". To study this question we examined the adherence to epithelial cells from trachea and from intestine
to bacteria of O78 strains carrying various pili. One such
example is presented in Fig. 4, where we compared cells
carrying AC/I pili and cells carrying K99 pili. The results
indicate that K99 pili, that are produced by cells involved

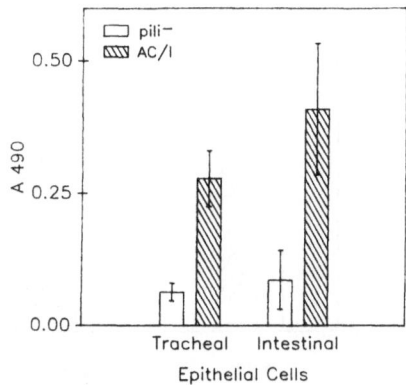

Fig. 3.   Adherence of *E. coli* O78 to epithelial cells. prep-
          aration of avian tracheal epithelial cell suspen-
          sion (8   x 10^7 cells/ml) was prepared   by scraping
          the inside of a trachea into phosphate- buffered
          saline (PBS) and washing. The bacterial suspension
          (2 x 10^9/ml) was prepared by scraping bacteria from
          nutrient agar   plates and washing them by centrifu-
          gation three   times and resuspending   in PBS.   The
          adhered bacteria were quantified with rabbit anti--
          O78 serum by ELISA using alkaline- phosphatase
          conjugates of goat-anti-rabbit serum, as described
          by Voller et al.(12).   Adherence of cells of strain
          781   grown at   18°C (not expressing AC/I pili) is
          shown by the blank box, grown at 37°C (expressing
          AC/I pili) by the striped box, and of unpiliated
          strain 278 grown at 37°C by the black box.

in intestinal infections,   adhere better to intestinal epithel, while cells carrying the AC/I pili, involved in extraintestinal infections initiated via the trachea do not show
the preferential adherence to the intestine.

Table 1. Adherence of O78 bacteria to avian and human epi-
          thelial cells.

Strain	Original Host	Pili	Adhering bacteria per 10^7 cells	
			human buccal	avian trachea
278	human	---	2.5 x 10^5	2 x 10^5
H10407	human	CFA/I	7.2 x 10^7	1.5 x 10^6
781	chicken	AC/I	1.0 x 10^6	4.8 x 10^7

The experiment was performed as described for Fig. 3. Buccal
cells were obtained by scraping and then as described for the
tracheal cell suspension. The number of adhering cells was
calculated from the ELISA data, using calibration curves.

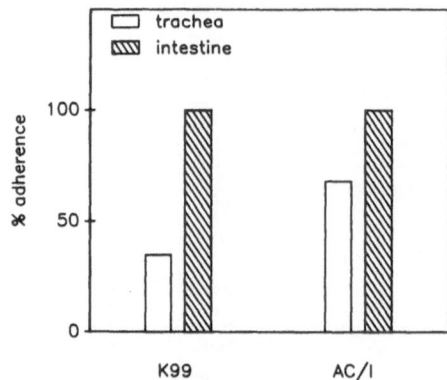

Fig. 4.   Adherence of *E. coli* O78 strains to tracheal and
intestinal epithel. Avian tracheal and intestinal
epithelial cells were obtained as described in Fig.
3. The cells were incubated with a suspension of
washed bacteria expressing AC/I pili (blank box) or
K99 pili (stripped box). The adhered bacteria were
quantified as in Fig. 3.

HOST SPECIFICITY *IN VIVO*

In order to examine the effect of the pili on bacterial-
host interaction *in vivo*, an experimental system was developed for inducing avian colisepticemia in a way similar to
the natural course of disease. In this model system chicks
are infected intra-tracheally and their weight is monitored
for about a week. The relative weight gain is an objective
measurement of morbidity, as it is slowed down by the disease. An example of such an experiment is presented in Fig.
5, and shows that weight gain is significantly lowered following infection with virulent bacteria. In this system we
have previously shown that O78 bacteria are more virulent
when expressing the AC/I pili (10).

This finding may simply reflects the need for adherence
pili in initiating the disease process, in which case bacteria carrying pili will be more virulent, independently of the
type of pili they express. On the other hand, it was possible
that AC/I pili are unique in intensifying the virulence of
avian *E. coli* O78 strains. To determine pili specificity *in
vivo* we performed the following experiment: chicks were
inoculated intra-tracheally with one of several strains of *E.
coli* O78 and their weight was monitored following inoculation. All the O78 strains used for inoculation carried a *colV*
plasmid, but expressed different pili.

The results presented in Fig. 6 show that strains
carrying AC/I pili are more virulent that strains carrying
other pili. This experiment indicates that there exists
host-specific virulence in avian strains of *E. coli* O78
expressing AC/I pili. However, at this stage it is not possible to conclude that this specificity depends <u>only</u> on the
presence of AC/I pili, as the avian strains were also more
virulent when administered intra-peritoneally (Fig. 6).

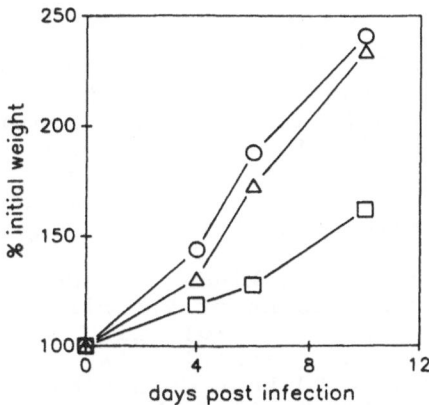

Fig. 5.   Effect of inoculation with *E. coli* O78 on weight of
chickens.   Three-week-old chickens were inoculated
intra-tracheally with 4 x 10⁷ cells of *E. coli* O78.

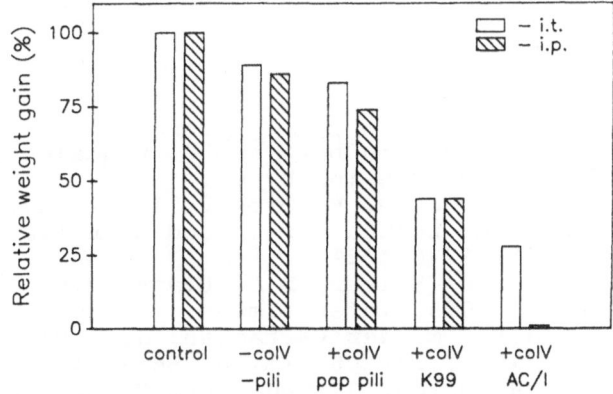

Fig. 6.   Effect of inoculation with various strains of *E.
coli* O78. Experiment was as in Fig. 5 except that
inoculation   was   intra- tracheally   with 4 x 10⁷
bacteria or   intra-peritoneally with 4 x 10⁵ bacteria.

        A detailed analysis of the virulence factors that play a
role in determining the host specificity of *E. coli* O78
'strains   will advance our   understanding of the complicated
host-bacteria interaction. In addition, the existence and
extent of host specificity of avian strains is important to
understand in view of the widespread passage of virulent *E.
coli* strains from one host to another.

SUMMARY

Escherichia coli strains of serotype O78 are virulent to mammals (including humans) and poultry, and are associated with local diseases (diarrhea, UTI, meningitis) or systemic diseases (septicemia). The broad host specificity of O78 strains, in addition to the diverse kinds of diseases that they cause, make it a good model system for studying the role of the various virulence factors in pathogenicity.

At least 5 different kinds of adherence pili are produced by the various strains of *E. coli* O78, and they include K99, K88, CFA/I, and P (20 kd) pili. In addition, strains associated with avian colisepticemia, produce pili with a molecular weight of 18 kd that do not hemagglutinate RBC. The presence of these pili (called by us avian coli #I "AC/I") results in an increased binding to chicken epithelial cells *in vitro* and to chicks' trachea *in vivo*.

In order to determine the importance of the type of adherence pili in bacteria-host interaction, we studied several *E. coli* O78 strains that had been isolated from human, chicks and lambs. All these strains were septicemic and carried the *colV* plasmid, but produced different adherence pili. These strains were compared using avian septicemia as a model system, in which we could show that mortality and morbidity correlated with the original host and the type of adherence pili produced.

REFERENCES

1. Jansson, P.-E., B. Lindberg and G. Widmalm. 1987. Carbohydrate Research 165:87-92.
2. Goldman, R., D. White, F. Orskov, I. Orskov, P.D. Rick, M.S. Lewis, A.K. Bhattacharjee and L. Leive. 1982. J. Bacteriol. 151:1210-1221.
3. Evans, D.G., R.P. Silver, D.J. Evans, D.G. Chase and S.L. Gorbach. 1975. Infect.Immun. 12:656-667.
4. Girardeau, J.P., M. der Vartanian, J.L. Ollier and M. Contrepois. 1988. Infect. Immun. 56:2180-2188.
5. Czirok, E., H. Milch, J. Madar and G. Semjen. 1977. Acta microbiol. Acad. Sci. hung. 24:115-126.
6. Milch, H., E. Czirok, J. Madar and G. Semjen. 1977. Acad. Sci. hung. 24:127-137.
7. Nagaraja, K.V., D.A. Emery, J.A. Newman and B.S. Pomeroy. 1982. Am.J.Vet.Res. 44:284-287.
8. Arp, L.H. and A.E. Jensen. 1980. Avian Dis. 24:153-161.
9. Dho, M. and J.P. Lafont. 1982. Avian Dis. 26:787-797.
10. Naveh, M.W., T. Zusman, E. Skutelsky and E.Z. Ron. 1984. Avian Dis. 28:651-661.
11. Yerushalmi, Z., N. Smorodinsky, M.N. Naveh & E.Z. Ron. 1990. Infect. Immun. 58:1129-1131.
12. Voller, A., A. Bartlett, & D.E. Bickwell. 1977. Clin. Chim. Acta 11:1-40.

THE ANALYSIS OF *BORDETELLA PERTUSSIS*

FIMBRIAL MUTANTS IN A RABBIT MODEL

Frits R. Mooi, Han G.J. van der Heide, Henk C.
Walvoort*, Henk Brunings, Wim H. Jansen, and Piet A.M.
Guinee

Department of Bacteriology and *Department of Control of
Bacterial Vaccines, National Institute of Health and
Environmental Protection, PO Box 1, 3720 BA Bilthoven
The Netherlands

INTRODUCTION

Bordetella pertussis is the causative agent of whooping
cough or pertussis, a serious respiratory disease (see ref. 1
for an excellent review). The first step in the pathogenesis
of pertussis is adherence of the bacteria to the respiratory
tract, and two classes of bacterial proteins are probably
involved in this process, filamentous hemagglutinin (FHA),
and fimbriae. After adherence, the bacteria proliferate and
produce a number of toxins: tracheal cytotoxin, dermonecrotic
toxin, adenyl cyclase and pertussis toxin. These toxins cause
local tissue damage, immune suppression, and are responsible
for the systemic effects of the disease. It has been suggested that pertussis toxin may also function as a colonization
factor, since it is able to bind to both the bacteria and the
host-cells (1).

B. pertussis adheres specifically to the cilia of respiratory epithelial cells. Analysis of transposon mutants has
revealed that FHA is implicated in this attachment (3). To
date, similar studies using well-defined fimbrial mutants
have not been performed. Therefore, the role of fimbriae in
the pathogenesis of pertussis is less clear. There is evidence that fimbriae are required for *B. pertussis* to infect
various animals (4,5), and fimbriae have been shown to be
protective antigens in the mouse intranasal model (6). Furthermore, there is also indirect evidence that fimbriae are
important for infection of humans (7). These observations are
consistent with the critical role which fimbriae play in the
establishment of a variety of bacterial infections (8).

At least two antigenically distinct fimbriae are produced
by *B. pertussis* (9,10). These structures were first identified as agglutinogens (i.e. antigens that induce antibodies
that cause agglutination of bacteria), and this has resulted
in some confusion with respect to nomenclature. According to

Microbial Surface Components and Toxins in Relation to Pathogenesis
Edited by E.Z. Ron and S. Rottem, Plenum Press, New York, 1991

69

Ashworth et al. (9), the two fimbriae correspond to the agglutinogens 2 and 3, while Cowell et al. (10) maintain that they correspond to agglutinogens 2 and 6. The serotype 2 and 3 fimbriae (as we will call them) are composed of subunits with molecular weights of 22,500 and 22,000, respectively (11,12). A single *B. pertussis* strain may produce only one fimbrial serotype, both serotypes, or no fimbriae at all. The mechanism underlying this phase variation has not yet been resolved. With a synthetic DNA probe, derived from a conserved part of the fimbrial subunit, at least three distinct fimbrial subunit genes were detected in *B. pertussis* strains (13). Two genes, designated *fim2* and *fim3*, code for the serotype 2 and 3 fimbrial subunits, respectively, while the third gene (*fimX*) codes for an as yet unidentified polypeptide. All three genes have been cloned and sequenced (14,15, 16). The derived amino acid sequences of the three subunits show extensive homology (approximately 60%). Homology is also observed with fimbrial subunits from *Escherichia coli* (17). Thus, *E.coli* and *Bordetella* fimbriae are evolutionary related.

Because adherence is the first step in the pathogenesis of pertussis, antibodies which block this process may be very effective in preventing disease. Therefore bacterial factors involved in adherence are potentially effective vaccine components. For this reason, our studies are focussed on the colonization factors of *B. pertussis*. Questions we are addressing are: Do FHA and fimbriae recognize the same or different receptors, are they required in different phases of infection, and if so, which phases. By studying these questions, we hope to determine whether or not these adhesins should be part of a future pertussis vaccine.

ANALYSIS OF *B. PERTUSSIS* MUTANTS IN A RABBIT MODEL

Using gene-replacement, we have constructed *fim* mutants (Table 1). An analysis of the constructed *fim* mutant B52 by means of immunoblotting revealed, that it was completely devoid of serotype 2 and 3 fimbrial subunits (Fig. 1). However, using polyclonal serum directed against the serotype 2 subunit, traces of a putative third fimbrial subunit were detected. We presume this fimbrial subunit is encoded by the *fimX* gene. Attempts to isolate the putative *fimX* product have failed, presumably because it is produced in very low amounts.

The *fim* mutant, a pertussis toxin mutant, a FHA mutant, and a wild-type strain were compared for their ability to colonize the rabbit (Fig. 2). The results of our rabbit experiments using the wild-type Tohama strain are consistent with those obtained by Preston et al.(4), who have shown that *B. pertussis* is able colonize the rabbit for five to ten months. A mutation in the pertussis toxin genes did not affect the ability of *B. pertussis* to infect the upper respiratory tract of the rabbit. Thus our results do not support a role for the pertussis toxin in adherence or persistence of *B. pertussis* in the rabbit. The observation that pertussis toxin is not required for colonization does not agree with a model proposed by Tuomanen (18), in which both FHA and pertussis toxin are required for adherence to ciliary cells.

Table 1. *B. pertussis* strains.

Strains	Relevant phenotype or genotype	Reference
Wellcome 28	serotype 2+, 3+	15
BP536 (Tohama)	serotype 2+, 3-; str	3
BP356	serotype *ptxs4*::tn5, str, kan	
	(derived from Tohama)	3
BP101	*fhaB*(delta101)	
	(derived from BP536)	3
B50	*fim2*::*SacI*, str	
	(derived from BP536)	▪
B52	*fim2*::*SacI*, *fim3*::kan, str, kan	
	(derived from BP536)	▪

▪Mooi et al., in preparation.

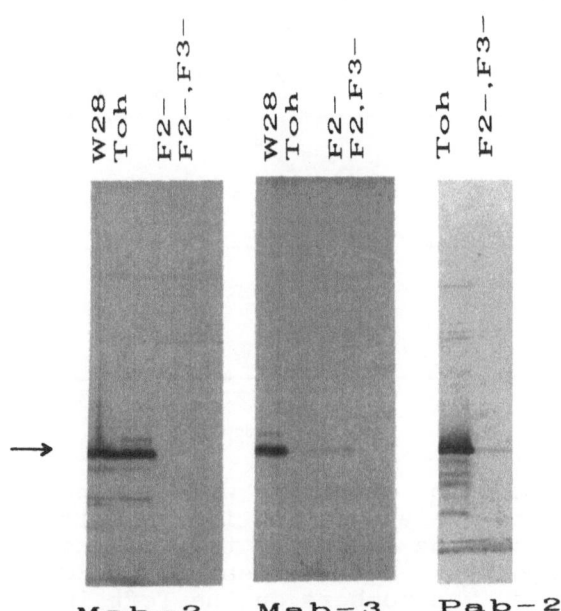

Fig. 1.　Immunoblots of *B. pertussis* strains. Total cell lysates were applied to the lanes. The arrow indicates the position of the fimbrial subunit. abbreviations: W28 = Wellcome 28 strain; Toh = Tohama strain BP536; F2- = *fim2* mutant strain B50; F2-,F3- = *fim2*, *fim3* mutant strain B52; Mab-2 = , monoclonal antibody directed against the *fim2* subunit; Mab-3 = monoclonal antibody directed against the *fim3* subunit; Pab-2 = polyclonal antibodies directed against the *fim2* subunit.

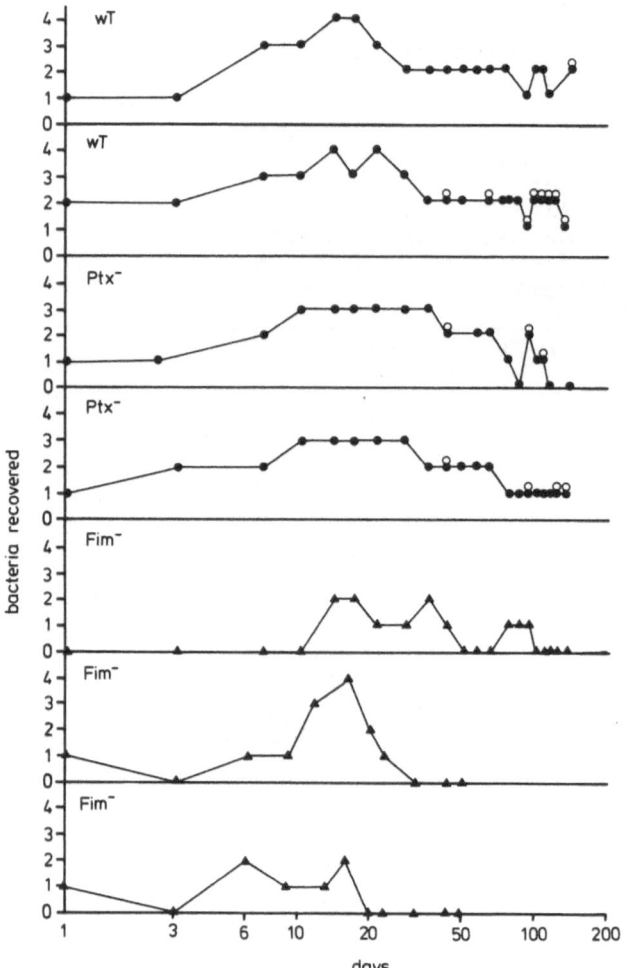

Fig. 2.   Infection of rabbits with *B. pertussis* strains. Rabbits were infected intranasally with 10⁶ bacteria. At the indicated time points, swabs from the nasopharynx were streaked on Bordet-Gengou agar plates. The numbers on the left indicate the number of bacteria isolated as described in Materials and Methods. The serotype of the strains is indicated by: ▲ (serotype 2-, 3-); ⊕ (serotype 2+, 3-); and ⊗ serotype 2+, 3+). Abbreviations: wt = wild-type strain B536; ptx- = pertussis toxin mutant strain BP356; fim- = *fim2, fim3* mutant strain B52.

The *fha* mutant was cleared from the rabbit within three days after infection (data not shown). This observation indicates that FHA plays a critical role early in the pathogenesis of pertussis, and is consistent with the finding that FHA is required for attachment to respiratory ciliary cells (2).

The *fim* mutant B52 was still able to colonize. However, in comparison to the wild-type strain and the pertussis toxin mutant, B52 bacteria were generally present in lower numbers, and sometimes eluded detection. Furthermore, the *fim* mutant was cleared more rapidly from the rabbit than the wild-type and pertussis toxin mutant strains. The *fim* mutant used in these studies has been mutated in the *fim2* and *fim3* genes. We have shown that *B. pertussis* contains at least one other fimbrial gene (*fimX*), and is still able to produce low amounts of a putative fimbrial subunit (Fig. 1). Thus the ability of the *fim* mutant strain to colonize, albeit inefficiently, may be due to the presence of (low amounts of) fimbriae on the cell surface encoded by *fimX* or other as yet undetected fimbrial genes.

At this point one may speculate that FHA is required in a very early step in the pathogenesis, for example the first attachment to a particular site within the respiratory tract. Fimbriae may function at a later stage, enhancing adherence or promoting spread to other locations of the respiratory tract. It is possible that fimbriae recognize receptors, different from the FHA receptors, which are not present at the primary attachment sites, but become accessible later in the disease due to tissue damage, or dissemination of the bacteria. Also, fimbriae may play a more important role late in infection, when the host immune response has become active, because in contrast to FHA, different fimbrial antigenic variants can be produced to evade the immune response.

FIMBRIAL PHASE VARIATION

The Wellcome 28 and the Tohama strains produce approximately equal amounts of serotype 2 fimbrial subunits as determined by immunoblotting (Fig. 1), and both strains agglutinate with serotype 2 antiserum. In contrast, compared to the Wellcome 28 strain, the Tohama strain produces very low amounts of serotype 3 fimbrial subunits (Fig.1). This difference in production is also indicated by agglutination, since in contrast to the Wellcome 28 strain, the Tohama strain does not agglutinate with serotype 3 antiserum. Thus the *fim3* gene is expressed at a very low level in the Tohama strain. We observed that in the rabbit the Tohama strain changed its serotype from 2+, 3- to 2+, 3+. The change in serotype is probably caused by immunological pressure, since it does not occur to this extent in vitro. This change in serotype was also observed with the pertussis toxin mutant, and has been described previously by Preston et al. (4). To study the mechanism underlying this phenomenon, we have cloned and sequenced the promoter regions of *fim3* genes derived from the Wellcome 28, and the Tohama strain. As discussed above, the former produces high amounts of *fim3* subunits, while the latter produces very low amounts of these polypeptides (Fig. 1). The sequences were compared with the putative promoter region of the *fim2* gene from the Wellcome

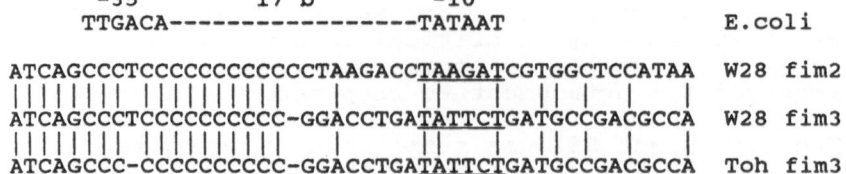

```
        -35           17 b          -10
        TTGACA-----------------TATAAT                    E.coli

ATCAGCCCTCCCCCCCCCCCCCCTAAGACCTAAGATCGTGGCTCCATAA    W28 fim2
|||||||| |||||||||  |   ||  |  ||  || | |
ATCAGCCCTCCCCCCCCCC-GGACCTGATATTCTGATGCCGACGCCA      W28 fim3
|||||||| ||||||||||  |    ||  || |  ||| | |  |
ATCAGCCC-CCCCCCCCCC-GGACCTGATATTCTGATGCCGACGCCA      Toh fim3
```

Fig. 3.   Putative fimbrial promoter sequences. The *E. coli*
           promoter consensus sequence is also shown. Tenta-
           tive - 10 sequences have been underlined. Bases
           conserved in all three fimbrial sequences have been
           indicated by |. Abbreviations: W28 = Wellcome 28 ;
           Toh = Tohama; b = bases.

28 strain (14)(Fig. 3). A striking feature of the three
promoter sequences is a C-stretch located at the presumed -35
position. The only difference observed between the fim3
promoter regions derived from the Wellcome 28 and the Tohama
strain was located in this C-stretch; a single T was absent
from the Tohama sequence. It has been shown that regions with
reiterated bases are hot spots for deletions and insertions,
due to transient misalignment during replication (19). A
similar mechanism may operate in phase variation of *B. pertussis* fimbrial genes. The insertions or deletions may affect
transcription, by changing the distance between DNA regions
involved in binding of RNA-polymerase and a positive regulator. Interestingly, it has been shown that the activity of
one of the *B. pertussis vir* gene products is regulated in a
similar manner. In this case regulation occurs at the level
of translation, since insertions or deletions in the C-
stretch affect the reading frame of a structural gene (20).

SUMMARY

    An essential first step in the pathogenesis of pertussis
is the adherence of *Bordetella pertussis* to the respiratory
tract. At least two classes of proteins are presumed to be
involved in this adherence, filamentous hemagglutinin (FHA)
and fimbriae. We have constructed well-defined fimbrial
mutants, and studied these mutants, together with FHA and
pertussis toxin mutants, in a rabbit model. In contrast to a
mutation in the pertussis toxin genes, mutations in the *fim*
or the *fha* genes affected the ability of *B. pertussis* to
colonize, and persist in the rabbit. Our results suggest that
inclusion of fimbriae and/or FHA in a vaccine may prevent
colonization by *B. pertussis*. Infection of rabbits with the
wild-type *B. pertussis* strain or the pertussis toxin mutant
resulted in changes in the fimbrial serotypes produced. We
present evidence that these changes are due to mutations in
the putative promoter regions of the *fim* genes.

ACKNOWLEDGEMENTS

    Part of this work was supported by a NATO travel grant

0527/87. Monoclonal antibodies were generously provided by dr J T Poolman, and B Kuipers. We thank I van Straaten-van de Kapelle, G Germans, H Gielen, and J F Visser for excellent technical assistance.

REFERENCES

1. Wardlaw A.C., and R. Parton. 1988. Pathogenesis and Immunity in Pertussis, John Wiley & Sons.
2. Tuomanen, E., A. Weiss, R. Rich, F. Zak and O. Zak. 1985. Proc Fourth Int Symp Pertussis. Dev Biol Stand 61:197-204.
3. Tuomanen, E., A. Weiss. 1985. J. Inf. Dis. 152:118-125.
4. Preston N.W., Timewell R.M., and Carter J.C. 1980. J. of Infection 2:227-235.
5. Stanbridge, T.N., and N.W. Preston. 1974. J. of Hygiene (Cambridge) 72:213.
6. Robinson A., L.A.E. Ashworth, A. Baskerville A., and L.I. Irons. 1985. Dev Biol Stand 61:165-172.
7. Preston N.W. 1988. In Pathogenesis and Immunity in Pertussis, Eds. Wardlaw A.C., and Parton R. John Wiley and Sons, 1-18.
8. De Graaf, F.K., and F.R. Mooi. 1986. In: Advances in Microbial Physiology, AH Roze and DW Tempest (eds). Academic Press, London, vol 28 pp:65-143.
9. Asworth, L.A.E., A. B. Dowsett, L.I. Irons, and A. Robinson. 1985. Proc. 4th Int. Symp. Pertussis, Geneva, 1984, Dev. Biol. Stand. 61:143-151.
10. Cowell J.L., J.M. Zhang, A. Urisu, A. Suzuki, A.C. Stevev, T. Liu, T.Y. Liu, C.R. Manclark. 1987. Infect. Immun. 55:916-922.
11. Irons L,I., L.A.E. Ashworth, A. Robinson. 1985. Proc. 4th Int. Symp. Pertussis, Geneva, 1984, Dev. Biol. Stand. 61:153-163.
12. Zhang J.M., J.L. Cowell, A.C. Steven, P.H. Carter, C.C. McGrath, and C.R. Manclark. 1985. Infect. Immun. 48:422-427.
13. Mooi F.R., H.G.J. van der Heide, A. ter Avest, K.G. Welinder, I. Livey, B.A.M. van der Zeijst, and W. Gaastra. 1987. Microbial Pathogenesis 2:473-484.
14. Livey I., C.J. Duggleby, and A. Robinson. 1987. Molecular Microbiology 1:203-209.
15. Mooi F.R., A.R. Relman, D. van Brenk, and S. Falkow. 1988. FEMS Symposium Pertussis, Berlin 1988, Book of abstracts.
16. Pedroni P., B. Riboli, F. de Ferra, G. Grandi, S. Toma, B. Arico and R. Rappuoli. 1988. Molecular Microbiology 2:539-543.
17. Mooi F.R. 1988. Ant. van Leeuwenhoek 54:465-474.
18. Tuomanen, E. 1988. In Pathogenesis and Immunity in Pertussis, Eds Wardlaw A.C., and Parton R. John Wiley & Sons, 75-94.
19. Streisinger, G., and J.E. Owen J.E. 1984. Genetics 109:633-659.
20. Stibbitz, S., W. Aaronson, D. Monack, and S. Falkow. 1989. Nature 338:266-269.
21. Sato, Y., H. Arai. 1972. Infect. Immun. 6:899-904.
22. Weiss, A.A., E.L. Hewlett, G.A. Myers, and S. Falkow. 1983. Infect. Immun. 42:33-41.

23. Relman, D.A., M. Dominighini, E. Tuomanen, R. Rappuoli and S. Falkow. 1989. Proc. Natl. Acad. Sci. USA 86 (in press).

DIFFERENTIATION BETWEEN PATHOGENIC AND

NON—PATHOGENIC *ENTAMOEBA HISTOLYTICA*

D. Mirelman

Macarthur Center for Molecular Biology of
Tropical Diseases and Department of Biophysics, Weizmann
Institute of Science, Rehovot, Israel

INTRODUCTION

The spectrum of disease in amebiasis is remarkable. It is estimated that there are approximately 500 million infections/year, but most of them are asymptomatic and only about 10% of the cases develop into disease (1,2). The prevailing hypothesis was that there are two distinct, albeit morphologically identical, species of *E. histolytica,* a pathogenic one which can cause disease and the second, a non pathogenic variety which is a commensal and will never cause disease (3). Numerous investigators are searching for biochemical and molecular differences between these two types of amoebae which may support this theory. Martinez-Palomo showed differences in agglutinability by the lectin Concanavalin A in *E. histolytica* isolates obtained from patients with invasive amebiasis and from symptomless cyst passers (4). Sargeaunt and his colleagues (5,6) have found that the two classes of amoebae can be differentiated by the electrophoretic migration of two isoenzymes, hexokinase and phospho glucomutase. The main questions that were raised, however, were (I) could persons infected with amoebae that have the non pathogenic isoenzyme patterns ever develop disease, (II) could asymptomatic carriers serve as the source of infection that would have clinical manifestations in other subjects, and (III) should asymptomatic carriers infected with amoebae that possess non pathogenic isoenzymes be subjected to drug therapy. Classification of *E.histolytica* into pathogenic or non pathogenic species by their isoenzyme electrophoretic patterns appears to be of no significant value because, as previously reported for other protozoa, notably Tetrahymena (7) and Paramecium (8), isoenzyme migration is not always a stable marker.

Changes in the isoenzyme pattern of cloned cultures of *E. histolytica* have been noted during the process of axenization and elimination of the accompanying bacterial flora (9,10). A more promising diagnostic assay, which depends on monoclonal antibodies which distinguish only strains with pathogenic isoenzyme patterns, was recently reported (11).

Microbial Surface Components and Toxins in Relation to Pathogenesis
Edited by E.Z. Ron and S. Rottem, Plenum Press, New York, 1991

Testing of strains which had switched their isoenzyme pattern from non pathogenic to pathogenic, however, showed that the antigen receptor for the monoclonal antibody appeared whenever the isoenzyme pattern changed.

In collaboration with William Petri from the University of Virginia, we have recently found interesting differences in the structure of the lectins from pathogenic and non pathogenic amoebae.. As previously found (12, 13) the Gal--inhibitable amebic lectins mediate adherence of theparasite to distinct receptors on mammalian cells and are a necessary step in the pathogenic process. The lectin from pathogenic strains has been purified and characterized (14). A number of monoclonal antibodies have been produced, but only a few of those cross react with the lectin of the non pathogenic strains. These preliminary results suggest that functional and perhaps topological differences may also exist between the two lectin molecules.

Recently we have also discovered chromosomal, circular-DNA, plasmid-like elements in trophozoites of *E. histolytica* (strain HM-1:IMSS) (15). Sequence analysis of segments of this plasmid revealed that it mostly comprises of ribosomal genes. The circular element contained, however, a non transcribing, intervening region consisting of 15 tandemly repeated units, about 145 bp each which were released by PvuI. Attempts to use these repetitive elements as DNA probes revealed that they gave a very distinct hybridization signal (at high stringent conditions) with strains that possessed pathogenic isoenzyme patterns, but not with isolates that had non pathogenic isoenzymes (16). Restriction maps and sequence analysis of the analogous extrachromosomal circular DNA from isolates with non pathogenic isoenzyme patterns showed that it also contained an area of tandem repeats, but these were of different size (133 bp) and sequence and were released by BamI. In parallel to what was observed with the P145 probe, the B133 probe did not hybridize to DNA of *E. histolytica* with pathogenic isoenzyme patterns. On the other hand, strains which had switched their isoenzyme pattern from non pathogenic to pathogenic pattern, hybridized exclusively with the P145 probe. Long exposures of southern blots containing DNA from non pathogenic trophozoites hybridized with P145 gave faint signals, suggesting that at least one or a few copies of such sequences may exist also in this other class of amoebae.

Our findings imply that *E. histolytica* trophozoites can apparently exist in two interconvertible states, and the switch from one state to the other can be triggered by drastic changes in their growth conditions, for example, elimination of the bacterial flora. The change from non-pathogenic state to the pathogenic one is apparently accompanied by numerous molecular and biochemical modifications. The mechanisms for such changes are not yet understood. One hypothesis which is currently under investigation is that depending on the environmental and growth conditions, differential amplification of either type of extrachromosomal element may occur in *E. histolytica* trophozoites. One type of extrachromosomal element may be produced and amplified under certain conditions, and when these change, a switch may occur causing the shutdown in the production of the first element and the amplification of the other. This would require that all *E. histolytica* should have an entire repertoire of genomic information whose replication and expression would be highly regulated. This could represent an ingenious mechanism to cope and adapt to changing conditions.

ACKNOWLEDGMENTS

The molecular biology studies described in the author's work were done in collaboration with Marion Huber, Leonard Garfinkel, Ann Chayen, Shmuel Rozenblatt, Carlos Gitler and Michel Revel and supported by a grant from the John D. and Catherine T. MacArthur Foundation. The lectin work is being carried out in collaboration with William Petri from the University of Virginia, Charlottesville.

REFERENCES

1. Martinez-Palomo, A. 1986. In Martinez-Palomo A . ed: "Amebiasis: Human Parasitic Diseases", Elsevier Biomed. Div. Amsterdam.
2. Walsh, J.A. 1988. In Ravdin J.I. ed: "Amebiasis: Human Infection by *Entamoeba histolytica*", John Wiley & Sons, New York, p. 93.
3. Brumpt, E. 1925. Bull de l'Acad Med. Paris 94:942.
4. Martinez-Palomo, A., Gonzales-Robles, A., de la Torre, M. 1973. Nature 245:186.
5. Sargeaunt, P.G., Williams, J.E. 1978. Trans R Soc Trop Med Hyg 72:164.
6. Sargeaunt, P.G. 1987. Parasitol Today 3:40.
7. Allen, S.L. 1968. Ann NY Acad Sci 151:190.
8. Rowe, E., Gibson, I., Cavill, A. 1970. Biochem Genetics 5:151.
9. Mirelman, D. et al.. 1986. Exp Parasitol 62:142.
10. Mirelman, D. et al.. 1986. Infect Immun 54:827.
11. Strachan, W.D. et al.. 1988. Lancet 12:561.
12. Mirelman, D., Ravdin, J.I. 1986. In: Microbial Lectins and Agglutinins: Properties and Biological Activity. D. Mirelman, ed. pp.319-334, John Wiley & Sons, N.Y.
13. Petri, W.A. et al.. 1987. J. Clin. Invest. 80:123.
14. Petri, W.A. et al. 1989. J. Biol. Chem. 264:3007.
15. Huber, M. et al. 1989. Mol. Biochem. Parasitol. 32:285.
16. Garfinkel et al. 1989. Infect. Immun. 57:926.

ADHERENCE AND PATHOGENICITY OF

MYCOPLASMA PNEUMONIAE: A REVIEW

Helmut Brunner

Institut für Chemotherapie, Bayer Pharma
Forschungszentrum, 5600 Wuppertal, FRG

INTRODUCTION

Mycoplasmas belong to the class Mollicutes, because they lack a rigid cell wall and do not possess precursors of peptidoglycan. They can rearrange their surface structure in direct response to an outside stimulus. Their limiting membrane is similar to the membrane of animal cells. They are the smallest free-living organisms able to grow on lifeless media.

There are more than 100 named Mycoplasma-species, 50 % of which have been associated with disease, underscoring their role as pathogens for men, animals and plants. Some species are normal inhabitants of the oropharynx and the genital tract of man and animals. The human and animal pathogens are extracellular parasites and colonize mucosal surfaces of the host. Three mycoplasma-species, *M. pneumoniae* (in men), *M. gallisepticum* (in chickens) and *M. pulmonis* (in mice and rats) are respiratory tract pathogens. All three are motile and adhere strongly to glass, plastic materials and host cells. Thus motility and adherence are of biological importance in the pathogenesis of the diseases due to these mycoplasmas. Three Mycoplasma species are human pathogens: *M. pneumoniae, M. hominis* and *Ureaplasma urealyticum*. Good evidence exists, that *M. genitalium*, a difficult to grow isolate from the genital tract and the oropharynx of man, is also involved in human disease. The major human pathogen, *M.pneumoniae*, has remained a consistent and significant cause of acute respiratory tract disease. Infection-rates range from 10-25 % in children and young adults with pneumonia (1,2,3,). The purpose of this paper is to review current knowledge on the host-pathogen-interaction of *M. pneumoniae* as a surface parasite of the human respiratory tract.

STRUCTURE OF *M. PNEUMONIAE*

M. pneumoniae shows an asymmetric structure. It possesses an organized tip, which consists of a central core, with a dense central filament (4,5,). The core is surrounded

Microbial Surface Components and Toxins in Relation to Pathogenesis
Edited by E.Z. Ron and S. Rottem, Plenum Press, New York, 1991

by a translucent space that is enveloped by an extension of the organism's plasma membrane. A 169 kilodalton protein (P1) is densely clustered in the membrane of the tip. This protein mediates the attachment of the organisms to glass, plastic and host tissue (5-11).

MOTILITY OF *M. PNEUMONIAE*

Bredt (12) has shown that *M. pneumoniae* is motile and that the tip with its dense patches of protein P1 is the leading head during motility. A cytoskeleton, found in *M. pneumoniae*, might facilitate the gliding motility of the organism (13). Motility and cytadsorption may be coordinated via the cyto-skeleton .This could require the presence of densely packed P1 on the tip.Kahane et al (14) have shown that P1 is associated with the cytoskeletal structure of *M. pneumoniae*.

SELF-ATTACHMENT OF *M. PNEUMONIAE*

During growth in artificial medium and presumably during colonization of the respiratory epithelium, *M. pneumoniae* forms highly entangled clumps, which have been called "speru-les" (15).They consist of several hundred organisms attached to each other.It is unclear, whether the tip is involved in self- attachment. The spherules may play a role in pathoge-nicity, because individual organisms in the center of a spherule are protected from host defense mechanisms, e.g. from the lytic action of antibodies and complement (16).

ATTACHMENT TO GLASS AND PLASTIC

Somerson et al (17) first demonstrated that *M. pneu-moniae* attaches to glass. The organisms form a confluent sheet which covers the entire surface of the bottles. The sheet can be removed by scraping or by treatment with tryp-sin. This discovery has greatly facilitated research on *M. pneumoniae*, because the organisms can be washed on the glass without physical stress by centrifugation. Taylor-Robinson and Manchee (11) demonstrated that *M. pneumoniae* also adheres to plastic. Feldner et al(18) presented evidence for the relationship between the attachment process of *M.pneumoniae* to glass and the utilization of metabolic energy.

ATTACHMENT OF *M. PNEUMONIAE* TO HOST TISSUE

Adherence of bacteria to animal cell surfaces is the most important step towards induction of disease. A two step parasitism occurs in *M. pneumoniae*. First *M. pneumoniae* attaches via the tip-like organelle to a sialic acid associ-ated receptor region (glycolipids) on the respiratory epithe-lium. The second step involves early abnormal host cell function with subsequent tissue cytopathology, when the attached virulent mycoplasmas have colonized the epithelium and continue their multiplication and metabolic activity. *M. pneumoniae* adheres to erythrocytes, leukocytes and respi-ratory epithelium of man and of several animal species.

Adherence of *M. pneumoniae* to erythrocytes has been studied very extensively, because tests on hemadsorption are easy to perform and cold hemagglutinins appear in the serum of 50% of patients with *M. pneumoniae* infection (9,19-25). Hemadsorption has first been described by Del Giudice and Pavia (19) who showed that guinea pig erythrocytes rapidly attach to *M. pneumoniae* colonies. Gorski and Bredt (20) obtained evidence that a protein was responsible for attachment to sheep erythrocytes and to serum-coated latex particles . Using scanning electron microscopy, Razin et al (9) studied the adherence of human erythrocytes to *M. pneumoniae* grown on glass and on plastic surfaces. The *M. pneumoniae* filaments frequently attached to the erythrocytes via their tip structure and distorted the shape of the erythrocytes by traction. The tight nature of the observed attachment of the microorganisms to the erythrocytes was interpreted as a possible indication of fusion of the erythrocyte membrane with the organisms at the point of attachment.

Attachment to erythrocytes in vitro and the severity of respiratory tract infections in animal models of disease are related, but a total absence of hemadsorption ability does not entirely eliminate the capability of *M. pneumoniae* to adhere to other host tissue (25-27). This means that the combining site for erythrocytes and those for other tissues are only partially identical. Kahane et al (22) detected P1-protein also in the hemadsorption-negative portion of virulent *M. pneumoniae*. P1 is densely clustered at the tip region of virulent *M. pneumoniae* but has also been shown in less-dense foci along the *M. pneumoniae* unit membrane (28,29).

Zucker-Franklin et al (30) studied the interaction of *M. pneumoniae* with peripheral blood leukocytes. The attraction of *M. pneumoniae* to the surface membranes of peripheral blood leukocytes and platelets was much more pronounced than the reactions between other bacteria and leukocytes. Powell et al (31) examined the interaction of *M. pneumoniae* with alveolar macrophages of guinea pigs by electron-microscopy. The microorganisms attached with their specialized terminal structure to the macrophage cell membrane. However, the attached organisms were not phagocytized until rabbit antimycoplasmal serum had been added.

Recently, the nucleotide sequence of the P1 attachment protein gene has been determined and the amino acid sequence deduced (32-34). The findings suggest, that the P1 gene is transcribed as part of a larger polycistronic message . The adherence protein has recently been isolated by Jacobs et al (35). *M. pneumoniae*-antigens have been expressed in *E. coli* (34).

M. genitalium is a newly isolated mycoplasma - species from the urethral specimens of human patients with non-gonococcal urethritis. Seroepidemiologic data and experimental infections suggest the pathogenicity of *M. genitalium* (36). Ultrastructural studies indicated that *M. genitalium* possesses a differentiated terminal structure covered by a peplomer-like nap (37-39).

Adherence of *M. genitalium* to Vero cell membranes appeared to be mediated by the nap area. This evidence suggests that *M. genitalium* may possess an attachment mechanism similar to that described earlier for *M. pneumoniae*. Homologies in the DNA and in the protein-sequence between the adhesins of *M. pneumoniae* and *M.genitalium* have recently been described (39).

TISSUE DAMAGE BY *M. PNEUMONIAE*

It has been speculated that membrane fusion might play some role in the attachment between mycoplasma cells and the membranes of host cells (9,40). Fusion between the tip of *M. pneumoniae* and erythrocyte membranes been been suggested but could not definitely be shown on the basis of scanning electron micrographs. Prakash and Gabridge (40) found that treatment with polyethylene glycol, a known fusogenic agent, increased the attachment of mycoplasmas to fibroblasts and mycoplasmas to each other. If fusion would occur, a variety of toxins could be introduced directly into the host cell. Upchurch and Gabridge (41) demonstrated that human lung fibroblasts infected with *M. pneumoniae* developed a cytopathic effect when the infected fibroblasts were in a state of rapid DNA and RNA metabolism suggesting a correlation between the ability of *M. pneumoniae* to produce cellular injury and the state of nucleic acid metabolism of the host cells. *M. peumoniae* induced oxidative damage by an increase in the intracellular levels of hydrogen peroxide and superoxide anions generated during *M. pneumoniae* infection (42). The organisms also inhibited host-cell catalase (43).

SHEDDING OF *M. PNEUMONIAE* AFTER CHEMOTHERAPY

The most important problem in chemotherapy of *M. pneumoniae* disease concerns the shedding of the organisms from the respiratory tract of man after chemotherapy, when the disease symptoms have been reduced (44). Studies in hamsters were undertaken to elucidate this rather interesting phenomenon, because experimental *M. pneumoniae* infection in the golden Syrian hamster resembles in several features atypical pneumonia in man caused by the same organisms (45,46). Figure 1 shows the rather long persistence of the organism in the respiratory tract of hamsters after intranasal inoculation of *M. pneumoniae*. Treatment with tetracycline, erythromycin or ciprofloxacin was able to reduce the histopathological lesions in the lungs of hamsters after intranasal inoculation of *M. pneumoniae*, whereas the number of colony-forming units (CFU) in the lung and in the upper respiratory tract remained high. This was especially surprising for erythromycin, which showed an extremely low in vitro minimal inhibitory concentration (MIC: 0.005 mg/l) for *M. pneumoniae*. Erythromycin, like tetracycline or ciprofloxacin, did not eradicate the organisms from the hamsters respiratory tract (45). These results confirm previous data obtained by Slotkin et al (46) for erythromycin and tetracycline. Resistant organisms during or after therapy were neither seen in our experiments nor in those of Slotkin et al.(45,46). The data closely reflect observations in man after therapy with erythromycin or tetracycline (44).

The persistent shedding of *M. pneumoniae* after therapy

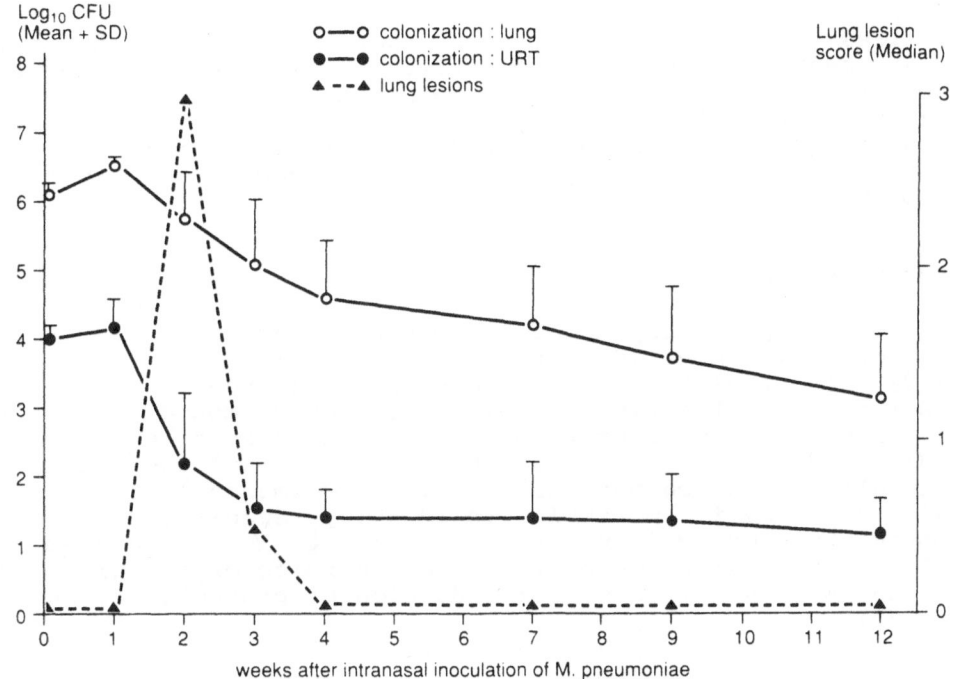

Fig. 1. Course of *M. pneumoniae* infection after intranasal inoculation of 10^6 colony forming units (CFU) of virulent organisms in hamsters. Persistent colonization of the lung and the upper respiratory tract are seen. The lung lesion score shows a maximum at two weeks after infection.

without development of antibiotic-resistant organisms, and the simultaneous beneficial clinical response may be explained in three ways: 1. The antibiotics inhibit replicating bacteria whereas non-multiplying organisms persist; 2. bacterial virulence properties are altered during or by the antibiotic treatment; 3. antibiotics inhibit bacterial growth until local host defense mechanisms have developed to a degree, at which attachment or other pathogenicity-mechanisms of the organisms are hampered.

ANTIMYCOPLASMAL AGENTS AND ATTACHMENT

It has been shown for *Streptococcus pyogenes* and *Escherichia coli,* that sublethal concentrations of various antibiotics markedly impaired adhesion of the bacteria to human cells (for review see Beachey et al, 47). Sandberg et al (48) have shown that antibodies to purified pili inhibited attachment of ampicillin-treated bacteria more efficiently than that of untreated bacteria. This indicates that the antibiotic unravelled attachment sites on the bacterial surface. Studies with sublethal concentrations of antibacterial agents, whose mode of action is well known, may elucidate genetic and chemical bacterial factors involved in colonization of the mucosa, because these drugs modify adherence. Hu et al (49) showed that trypsin-treated mycoplasmas which had lost P1 and other membrane proteins restored attachment capa-

bilities and regained P1, when they were reincubated into
fresh medium. Addition of erythromycin during the re-
incubation suppressed the regeneration of P1 and prevented
reattachment of *M. pneumoniae* to respiratory epithelium.The
data support the biosynthetic nature of P1 and ruled out the
possibility that this protein is adsorbed or incorporated
from the medium.

IMMUNE RESPONSE

Glycolipid antigens of *M. pneumoniae* dominated serologic
studies until about 1982, when protein antigens were isolated
(7,8). The isolation of protein antigens, the evidence for
their role in attachment, the production of monoclonal antibodies to attachment proteins and the demonstration of antibodies to these proteins in the sera and nasal secretions of
naturally infected persons and in experimentally inoculated
animals makes it possible to identify specific pathogenic
components involved in infection due to *M. pneumoniae* and
will permit evaluation of the relative roles of the glycolipid and protein antigens for induction of protective immunity
(50-52).

LOCAL ANTIBODIES

Immunization with the P1-protein may provide a highly
selective form of protective immunity. However, it will
probably be necessary to induce local antibodies in the
respiratory tract for significant protection (53,54). It must
also be excluded that any vaccine leads to sensitization and
results in more severe disease after infection with virulent
organisms than in unvaccinated persons.

In animal experiments, high numbers of virulent *M. pneumoniae*-organisms exposed for 4 h at 37°C to monoclonal antibodies (MOAB) directed against the attachment-tip were inoculated intranasally into hamsters. A significant reduction in
the lung lesion score, but not in the numbers of organisms in
lung tissue or wash- fluid of the upper respiratory tract,
was seen in hamsters 14 days after inoculation of MOAB--
treated organisms as compared to controls (55). Further
studies are needed to show whether antibodies to the tip
will enhance the eradication of the organisms from the respiratory tract.

REPEATED ENCOUNTER WITH THE ORGANISMS AND LUNG HISTO-
PATHOLOGY

A role of allergic reactions is postulated for disease
caused by *M. pneumoniae* (56). Infection with *M. pneumoniae* is
common in children less than five years of age, but generally
does not produce clinical disease in this age group. It has
thus been hypothesized that the pathogenesis of pneumonia due
to this organism may at least partly involve an immunopatho-
logical process and the severity of lung lesions depend on
early sensitization by inapparent infections. The presence of
erythema nodosum and the occurrence of the Stevens Johnson
Syndrome are suggestive of allergy. Individuals vaccinated

86

with an inactivated vaccine who did not develop growth inhibiting antibodies were more severely affected on challenge
than non-vaccinated subjects (44). Nevertheless, sensitization in the pathogenesis of *M. pneumoniae* disease is still a
controversial issue.

CROSS REACTION WITH HOST TISSUE

M. pneumoniae disease in man is commonly associated
with the transient production of autoantibodies to a carbohydrate antigen(I), which occurs on erythrocytes and other
tissues (57). *M. pneumoniae* infection is therefore an excellent natural model for studying host-parasite interactions
and the pathogenesis of autoimmunity in man.

Investigations by Loveless and Feizi (58) on the erythrocyte receptors for this agent have shown that they involve
long-chain sialo-oligosaccharides of I and i antigen type.
Antigens of this family have also been shown to occur on
human lymphocytes, monocytes, and granulocytes.Among the
mature epithelial cells, only ciliated cells were found to
express the long-chain antigens, whereas mucus-secreting
cells contained the short-chain antigens associated with
mucus globules. These findings have led to the proposal that
in the inflamed areas of the respiratory tissues, receptor
-mediated complexes formed between the host oligosaccharides
and the lipid-rich mycoplasma (functioning as an adjuvant)
may serve as a trigger for autoimmunization. Moreover, the
occurrence of the receptor structures on monocytes, which are
accessory cells with important roles in antigen presentation
may potentiate the immunogenic stimulus and account for the
frequent occurrence of the autoimmune response after infection.

REFERENCES

1.   Foy, H.M., Kenny, G.E., Cooney, M.K. and Allan, I.D.
     1979. J. Infect. Dis. 139:681-687.
2.   Lind, K. and Bentzon, M.W. 1976. Int. J. Epidemiol.
     5:267-277.
3.   Murphy, T.F., Henderson, F.W., Clyde, W.A., Collier,
     A.M. and Denny, F.W. 1981. Am. J. Epidemiol. 113:12-21.
4.   Biberfeld, G. and Biberfeld, P. 1970. J. Bacteriol.
     102:855-861
5.   Collier, A.M. 1983. Rev. Infect. Dis. 5:685-691.
6.   Carson, J.L., Collier, A.M. and Hu, S-CS. 1980. Infect.
     Immun. 29:1117-1124.
7.   Feldner, J., Goebel, U. and Bredt, W. 1982. Nature
     298:765-766.
8.   Hu, P.C., Cole, R.M., Huang, Y.S., Graham, J.A., Gard-
     ner, D.E., Collier, A.M. and Clyde,Jr. W.A. 1982.
     Science 216:313-315.
9.   Razin, S., Banai, M., Gamliel, H., Polliack, A.,
     Bredt, W. and Kahane, I. 1980. Infect. Immun. 30:
     538-546.
10. Sobeslavsky, O., Prescott, B. and Chanock, R.M. 1968.
     J.Bacteriol. 96:695-705.
11. Taylor-Robinson, D. and Manchee, R.J. 1967. J. Bacte-
     riol. 94: 1781-1782.

12. Bredt, W. 1979. In: Barile, M.F., Razin, S. eds.. The mycoplasmas. Academic Press, New York, vol. 1: 141-155.
13. Neimark, H.C. 1977. Proc. Natl. Acad. Sci. USA, 74:4041-4045.
14. Kahane, I., Tucker, S., Leith, D.K., Morrison -Plummer, J. and Baseman, J.B. 1985. Infect. Immun. 50: 944-946.
15. Boatman, E.S. and Kenny, G.E. 1971. J. Bacteriol. 106:1005-1015.
16. Brunner, H., Razin, S., Kalica, A.R. and Chanock, R.M. 1971. J. Immunol. 106:907-915.
17. Somerson, N.L., James, W.D., Walls, B.E. and Chanock R.M. 1967. Ann. N.Y. Acad. Sci. 143: 384-389.
18. Feldner, J., Bredt, W. and Razin, S. 1981. Infect. Immun., 31:107-113.
19. Del Giudice, R.A. and Pavia, R. 1964. In: Proceedings of the Annual Meeting of the American Society for Microbiology, 71.
20. Gorski, F. and Bredt, W. 1977. FEMS Microbiology Letters 1:265-267.
21. Hansen, E.J., Wilson, R.M. and Baseman J.B. 1979. Infect. Immun. 23:903-906.
22. Kahane, I., Tucker, S. and Baseman, J.B. 1985. Infect. Immun. 49:457-458.
23. Denny, F.W., Clyde, W.A. and Glezen, W.P. 1971. J. Infect. Dis. 123:74-92.
24. Lipman, R.P. and Clyde, Jr., W.A. 1969. Proc. Soc. Exp. Biol. Med. 131:1163-1167.
25. Brunner, H., Greenberg, H., James, W.D., Horswood, R.L. and Chanock, R.M. 1973. Ann. N.Y. Acad. Sci. 225:436-452.
26. Feldner, J. and Bredt, W. 1983. J. Gen. Microbiol. 129:841-848.
27. Lipman, R.P., Clyde,Jr., W.A. and Denny, F.W. 1969. J. Bacteriol. 100:1037-1043.
28. Baseman, J.B., Cole, R.M., Krause, D.C. and Leith, D.K. 1982. J. Bacteriol. 151:1514-1522.
29. Brunner, H., Krauss, H., Schaar, H. and Schiefer, H.G. 1979. Infect. Immun. 24:906-911.
30. Zucker-Franklin, D., Davidson, M. and Thomas, L. 1966. J. Exp. Med. 124:521-531.
31. Powell, D.A., Hu, P.C., Wilson, M., Collier, A.M. and Baseman, J.B. 1976. Infect. Immun. 13:956-966.
32. Dallo, S.R., Su, C-J., Horton, J.R. and Baseman, J.B. 1988. J. Exp. Med. 167:718-723.
33. Inamine, J.M., Denny, T.P., Loechel, S., Schaper, U., Huang, C.H., Bott, K.F. and Hu, P.C. 1988. Gene 64:217-229.
34. Trevino, L.B. Haldenwang, W.G. and Baseman, J.B. 1986. Infect. Immun. 53:129-134.
35. Jacobs, E., Fuchte, K. and Bredt, W. 1988. Biol. Chem. Hoppe-Seyler 369:1295-1299.
36. Moller, B.R., Taylor-Robinson, D. and Furr, P.M. 1984. Lancet i:1102-1103.
37. Morrison-Plummer, J., Lazzell, A. and Baseman, J.B. 1987. Infect. Immun. 55:49-56.
38. Dallo, S.F., Chavoya, A., Su, C-J. and Baseman, J.B. 1989. Infect. Immun. 57:1059-1065.
39. Yogev, D. and Razin, S. 1986. Int. J. Syst. Bacteriol. 36:426-430.

40. Prakash, G. and Gabridge, M.G. 1981. Infect. Immun. 32:969-972.
41. Upchurch, S. and Gabridge, M.G. 1981. Infect. Immun. 31:174-181.
42. Almagor, M., Kahane, I., Gilon, C. and Yatziv, S. 1986. Infect. Immun. 52:240-244.
43. Almagor, M., Yatziv, S. and Kahane, I. 1983. Infect. Immun. 41:251-256.
44. Smith, C.B., Friedewald, W.T. and Chanock, R.M. 1967. J. Am. Med. Assoc. 199:353-358.
45. Brunner, H. and Zeiler, H.J. 1986. pp 77-81. In: Neu, H.C.,and Weuta, W.,eds.., Proceedings of the 1st International ciprofloxacin workshop , Excerpta Medica 1986.
46. Slotkin, R.I., Clyde, W.A. and Denny, F.W. 1967. Am. J. Epidemiol. 86:225-237.
47. Beachy, E.H., Eisenstein, B.I. and Ofek, I. 1981. In: Adhesion and microorganism pathogenicity. Ciba Foundation symposium 80, p. 288-305. Pitman Medical, Tunbridge Wells.
48. Sandberg, T., Stenquist, K. and Svanborg Eden, C. 1979. Rev. Infect. Dis. 1:838-844.
49. Hu, P.C., Collier, A.M. and Baseman, J.B. 1977. J. Exp. Med. 145:1328-1343.
50. Hu, P.C., Huang, Y-S., Graham, J.A. and Gardner D.E. 1981. Biochem. Biophys. Res. Comm. 103: 1363-1370.
51. Jacobs, E., Stuhlert, A., Drews, M., Pumpe, K., Schaefer, H.E., Kist, M. and Bredt, W. 1988. Microbial Pathogenesis 5:259-265.
52. Leith, D.K., Trevino, L., Tully, J.G., Senterfit, L.B. and Baseman, J.B. 1983. J. Exp. Med. 157: 502-514.
53. Brunner, H., Greenberg, H.B., James, W.D., Horswood, R.L., Couch, R.B. and Chanock R.M. 1973. Infect. Immun. 8:612-620.
54. Loos, M., and Brunner, H. 1979. Infect. Immun. 25: 583-585.
55. Brunner, H., Feldner, J. and Bredt, W. 1984. Israel J. Med. Sci. 20:878-881.
56. Brunner, H., Prescott, B., Greenberg, H., James, W.D., Horswood, R.L. and Chanock, R.M. 1977. J. Infect. Dis. 135:524-530.
57. Loomes, L.M., Uemura, K.I., Childs, R.A., Paulson, J.C., Rogers, G.N., Scudder, P.R., Michalski, J.C., Hounsell, E.F., Taylor-Robinson, D. and Feizi, T. 1984. Nature London. 307:560-563.
58. Loveless, R.W. and Feizi, T. 1989. Infect. Immun. 57:1285-1289.

II. CELL INVASION AND INTRACELLULAR MULTIPLICATION

GENETICS OF FACTORS INVOLVED IN EXTRAINTESTINAL

E. COLI PATHOGENICITY

Jörg Hacker[1], Reinhard Marre[2], Joachim Morschhäuser[1],
Manfred Ott[1], and Thomas Schmoll[1]

[1]Insitut für Genetik und Mikrobiologie, Univ.
Würzburg, Röntgenring 11, D- 8700 Würzburg and
[2]Institut für Medizinische Mikrobiologie,
Ratzeburger Allee 160, Univ. Lübeck, D- 2400 Lübeck

PATHOGENICITY FACTORS OF EXTRAINTESTINAL *E. COLI* STRAINS

Escherichia coli strains may cause urinary tract infections (UTI), sepsis and cases of new born meningitis (NBM).
It has been shown that different factors contribute to pathogenicity of extraintestinal *E. coli* strains. Certain types
of O antigen (01, 02, 018, 075) and capsule (K1, K5, K15) are
necessary for resistance of pathogens to the action of the
immune system especially to lysis of serum complement compounds (1). Siderophore systems like aerobactin molecules are
able to take up iron from the environment of the strains and
they support extraintestinal *E. coli* with these ions (2).
Hemolysins represent protein toxins which are able to disrupt
eukaryotic cells (3). Under sublethal concentrations hemolysin proteins stimulate the release of mediators of inflammation from leukocytes and mast cells (4).

Adhesins which are often associated with the occurrence
of fimbriae enable bacteria to attach to eukaryotic tissues
and erythrocytes. Fimbrial adhesins of *E. coli* can be distinguished by their binding specificity. P fimbriae which are
produced by the majority of uropathogenic strains are able to
bind to the alpha-D-Gal-beta-(1-4) D-Gal region of receptor
structures. Type I fimbriae are commonly distributed among
pathogenic and non-pathogenic *E. coli* strains and they interact with mannose-containing receptors (5). S fimbriae which
are found among NBM strains and, to a minor extend among UTI
isolates recognize alpha-Sialic acid-(2-3)-beta-Gal containing receptor molecules (6).

GENETIC ANALYSIS OF THE S FIMBRIAL ADHESIN (*sfa*) GENE CLUSTER

S fimbriae are produced by NBM strains of serotypes
02:K1, 018:K1 and 083:K1 as well as by UTI isolates of the O

Microbial Surface Components and Toxins in Relation to Pathogenesis
Edited by E.Z. Ron and S. Rottem, Plenum Press, New York, 1991

93

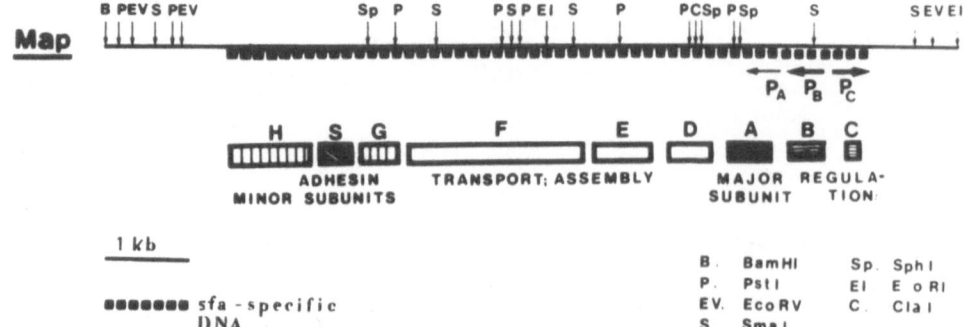

Fig. 1.   Genetic structure of the S fimbrial adhesin (*sfa*) determinant. The recognition sites for different restriction enzymes and the promoter regions are indicated. The boxes represent the *sfa* specific genes (for further explanation see text).

serogroup O6 (7). We have cloned the S fimbrial adhesin (*sfa*) determinants from the chromosomes of O18:K1 and O6:K15 strains (8, Zirngibl and Hacker, unpublished). The *sfa* gene cluster of the O6 strain 536 was subcloned and fully sequenced. As indicated in Fig.1 about 7 kb of DNA are necessary for the determination of the S fimbrial adhesin. From the nine genes of the determinant four loci, *sfa*A, *sfa*G, *sfa*S and *sfa*H code for fimbrillin subunits, proteins of 16 kilodalton (kd), 17 kd, 14 kd and 32, respectively. One of these fimbrillins, *fa*A, a protein of 16 kd is identical to the major subunit but it does not mediate the S specific binding property (9). In order to determine the gene which codes for the S specific adhesin mutations were introduced into the genes *sfa*G, *sfa*S and *sfa*H. Clones carrying mutations in *sfa*G and *sfa*H were still able to agglutine erythrocytes. Mutations in the gene *sfa*S, however, completely eliminated binding. It was therefore concluded that the gene *sfa*S codes for the S specific adhesin.

The identity of the *fa*S minor subunit protein and the S specific adhesin was further elucidated by DNA- and protein sequence studies. The N-terminal amino acid sequence of the isolated protein *fa*S which was able to agglutinate erythrocytes was identical to the N-terminal protein sequence deduced from the DNA sequence of the gene *sfa*S (10, 11). The amino acid sequence of *fa*S is given in Fig.2. *fa*S carries a leader peptide of 22 amino acids. The mature S adhesin consists of 141 amino acids. It carries two cystine residues at well defined positions (aa +16 and aa +53) and a tyrosine residue at position +140 which is typical for all subunit proteins known (12). The N- and C-termini of the adhesin show similarities to the major subunit protein *fa*A suggesting a function of these regions in the interaction of the fimbrillin subunits during fimbria formation. In order to determine the putative Sialic acid binding site of the S adhesin the *sfa*S sequence was compared to protein sequences of other Sialic acid binding proteins like the K99 adhesin, the colinization factor antigen (CFA) II, and the B subunits of the cholera as well as the *E. coli* LT toxin (data from ref. 13). It was evident that sequence homologies existed between

-22   -1
MetLysLeuLysAlaIleIleLeuAlaThrGlyLeuIleAsnCysIleAlaPheSerAlaGlnAla
+1   +25
ValAspThrThrIleThrValThrGlyAsnValLeuGlnArgThrCysAsnValProGlyAsnValAspValSer
                                                                      +50
LeuGlyAsnLeuTyrValSerAspPheProAsnAlaGlySerGlySerProTrpValAsnPheAspLeuSerLeu
                                                                      +75
ThrGlyCysGlnAsnMetAsnThrValArgAlaThrPheSerGlyThrAlaAspGlyGlnThrTyrTyrAlaAsn
                                                                      +100
ThrGlyAsnAlaGlyGlyIleLysIleGluIleGlnAspArgAspGlySerAsnAlaSerTyrHisAsnGlyMet
                                                                      +125
PheLysThrLeuAsnValGlnAsnAsnAsnAlaThrPheAsnLeuLysAlaArgAlaValSerLysGlyGlnVal
                                  +141
ThrProGlyAsnIleSerSerValIleThrValThrTyrThrTyrAla*

Fig. 2.   Amino acid sequence of the S specific adhesin *fa*S.
           The positively charged amino acids are indicated by
           boxes.

epitopes of these proteins which had been shown to be involved in Sialic acid binding and a *fa* specific epitope
located between the two lysine residues at positions +116 and
+122 (see Fig.2). Three of the seven amino acids of this
epitope of *fa*S are positively charged. It can therefore be
speculated that the epitope between aa +116 and aa +122 may
be part of the binding site of *fa*S involved in the interaction of this protein and the negatively charged Sialic acid--
part of the receptor molecule.

The three genes *sfa*D, *sfa*E and *sfa*F of the *sfa* determinant code for proteins which are involved in the transport of
the major and minor subunits across the outer membrane and in
assembly of the fimbrial rod (Schmoll and Hacker, unpublished). Another two genes, *sfa*B and *sfa*C located at the imme-
diate 5` end of the determinant play a role in the regulation
of transcription (14, Schmoll and Hacker, unpublished).
Isolating gene fusions between *sfa* and the indicator genes
lacZ and *phoA*, coding for beta- galactosidase and alkaline
phosphatase, respectively, three promoters, PA, PB and PC
were identified next to the 5` end of the determinant (see
Fig.1). The existence of the promoters was confirmed by
primer extension and sequence studies (15, Morschhauser and
Hacker, in preparation).

ENVIRONMENTAL CONDITIONS INFLUENCING THE EXPRESSION OF S
FIMBRIAE

It has been shown that the expression of virulence
determinants of different species depends highly on the
environmental conditions of the surrounding milieu (16). In
order to determine the signals important for the expression
of S fimbriae in a wild-type pathogen the *sfa* promoter region
was fused to the *lacZ* indicator gene. Using the pir-dependent
vector system pRT733, developed by Mekalanos and co-workers
(16), the *sfa* -*lacZ* fusion was site specifically introduced
into the chromosome of a LacZ negative mutant of the uropathogenic O6 strain 536 resulting in the strain 536-9B4/12.
The beta-galactosidase values were equivalent to the expression of *sfa* under certain conditions (17). It was shown that
the *sfa*-lacZ wild-type hybrid strain 536/9B4-12 is able to
produce about 800 LacZ values under optimal conditions.

Table 1. Influence of environmental conditions on the expression of the S fimbrial adhesin determinant: Use of *sfa- lacZ* wild-type fusion of strain 536

Environmental conditions	Optimal expression[1]	Low expression[2]
Growth medium	Solid	Liquid
Growth phase	Early stationary phase	Exponential phase
Content of glucose	<0.5%	>0.5%
Temperature	37°C	<35°C; >39°C
Content of NaCl	0.5 %	<0.5%; >0.5%;
Oxygen	aerobic	anaerobic

[1] 600 - 800 LacZ units produced by strain 536-9B4/12 (*sfa*-lacZ)
[2] <200 LacZ units produced by strain 536-9B4/12 (*sfa*-lacZ)

In Table 1 the conditions necessary for maximal expression of *sfa* are given. Higher levels of *sfa* determinant expression occurs when cells are grown on solid media rather than in liquid culture. After 12 hours of growth the level of LacZ reached values of nearly 800 units. The presence of glucose strongly inhibited expression of the *sfa* determinant. In addition, the expression of *sfa* is strongly favoured by temperature of 37°C, 0.5% NaCl and aerobic conditions. Thus different environmental factors contribute to the expression of *sfa* in a wild-type pathogen.

ROLE OF S FIMBRIAE DURING INFECTION

In order to answer the question whether or not the S fimbrial adhesin plays a role during infection *sfa*-negative mutants were selected from the *sfa*-positive wild-type O6 strain 536 by transposon mutagenesis. One of these mutants, 536/17B1, carries a Tn*phoA* element inserted into the gene *sfa*A coding for the S fimbrial major subunit (see above). The isogenic strains which only differ in the expression of S fimbriae were tested in a rat pyelonephritis model (18). Seven days after intraurethral injection of 5x10[7] bacteria into rats the number of colonies per gram kidney were counted. As shown in Table 2 the bacterial counts decreased drastically in the *sfa*-negative strain 536/17B1 when compared to the *sfa*-positive isolate 536 WT. This indicates that *sfa* plays an important role in the pathogenicity of strain 536 in the rat pyelonephritis model.

In addition, different recombinant DNAs were introduced

into a *sfa*-negative variant of strain 536. This strain 536-21 which was isolated as a spontaneous mutant of strain 536 had also lost the ability to produce hemolysin (Hly) and P-related fimbriae (Prf) and also the serum resistance property (Sre) (19, 20). One of the plasmids used, pANN801-13, codes for S fimbrial production and mediates S specific binding (*sfa*A+, *sfa*S+). The plasmids pANN801-1 and pANN801-13/Tn5-026, however, code for S specific binding without fimbrial formation (*sfa*A-, *sfa*S+) and for S fimbria without binding specificity (*sfa*A+, *sfa*S-), respectively. As shown in Table 2 only strains carrying the plasmid pANN801-13 which code for the whole S fimbrial adhesin increased the nephro-pathogenicity of mutant 536-21 by a factor of five, the strains expressing fimbriae without binding or S specific adherence without fimbria formation had no influence on the pathogenicity of strain 536-21.

VARIATION OF PATHOGENICITY PATTERN DURING INFECTION

As mentioned above the strain 536-21 had lost several virulence properties such as the production of adhesins and hemolysin and serum resistance. The variant 536-21 was isolated in the laboratory. Genetic analysis of this strain made it clear that chromosomal deletions which had an influence on the pathogenicity pattern of this strain had occurred (19). It was demonstrated recently that the determinants coding for hemolysin (*hly*) and P-related (*prf*) were deleted in strain 536-21 whereas the *sfa* determinant, while still present on

Table 2. Characterization of *E. coli* wild-type clones expressing different *sfa* phenotypes

Strain	Phenotypes[1]					Nephrovirulence CFU/g kidney [2]
	faA	faS	Prf	Hly	Sre	
536WT	+	+	+	+	+	9.1×10^4
536/17B1	-	-	+	+	+	1.3×10^4
536-21	-	-	-	-	-	7.1×10^2
536-21 pANN801-1	-	+	-	-	-	7.0×10^2
536-21 pANN801-13/ Tn5-026	+	-	-	-	-	8.0×10^2
536-21 pANN801-13	+	+	-	-	-	3.0×10^3

[1] Abbreviations: faA, production of S fimbriae; faS, production of S specific adhesins; Prf, production of P related fimbriae; Hly, hemolysin production; Sre, serum resistance
[2] Colony forming units per gram kidney seven days after intraurethral injection of 5×10^7 bacteria into rats

Table 3. Genetic characterization of wild-type strain 536 and mutants.

Strain	Origin	Gene determinants[1]			
		sfa	*prf*	*hlyI*	*hlyII*

Strain	Origin	*sfa*	*prf*	*hlyI*	*hlyII*
536	uropathogenic wild-type	present expressed	present expressed	present expressed	present expressed
536-21	laboratory mutant	present repressed	deleted	deleted	deleted
536-22	*in vivo* generated mutant	present repressed	deleted	deleted	deleted

----

[1] for abbreviations see Table 2

the chromosome, was not expressed (see Table 3). The DNA regions which are absent in the mutant strain may therefore additionally code for a factor which positively influences the expression of S fimbriae in the wild- type strain.

To see whether or not such deletions also occur *in vivo* bacteria were isolated from the kidneys of rats at various times after intraurethral injection. The isolated bacteria were tested for virulence. Seven days after infection one out of 40.000 colonies was selected which showed the same properties as strain 536-21 (Table 3). The DNA of this mutant strain, termed 536-22 was isolated, cleaved with the restriction enzyme *XbaI* and analyzed by pulse-field gel electrophoresis. It is shown in Fig.3 that the XbaI-cleavage pattern of chromosomal DNAs isolated from strain 536-22 differs from the pattern specific for the wild-type strain 536 (lanes E, F, and H). The observed differences argue for chromosomal rearrangements and deletions in the genome of 536-22 compared to the 536 genome.

It is also evident from Fig. 3 that the *sfa* gene cluster was still present in the chromosome of strain 536-22. It is interesting to see that the *sfa* determinant is located on *XbaI* fragments of identical sizes in the genomes of the O6:K15 strains as well as in those of the O18:K1 isolates analyzed here (lanes c-f and h) while the DNA of a O2:K1 isolate shows a different pattern (lane b). In contrast to the *sfa* determinant which was still present in the 536-22 genome the *hly* and *prf* sequences were deleted as already shown for the laboratory mutant 536-21 (see Table 3). Therefore it can be concluded that non-pathogenic variants of wild-type strains due to chromosomal deletions are not restricted to the laboratory and that they also originate *in vivo*.

SUMMARY

Escherichia coli strains are able to cause urinary tract infections, sepsis and new born meningitis. Different factors

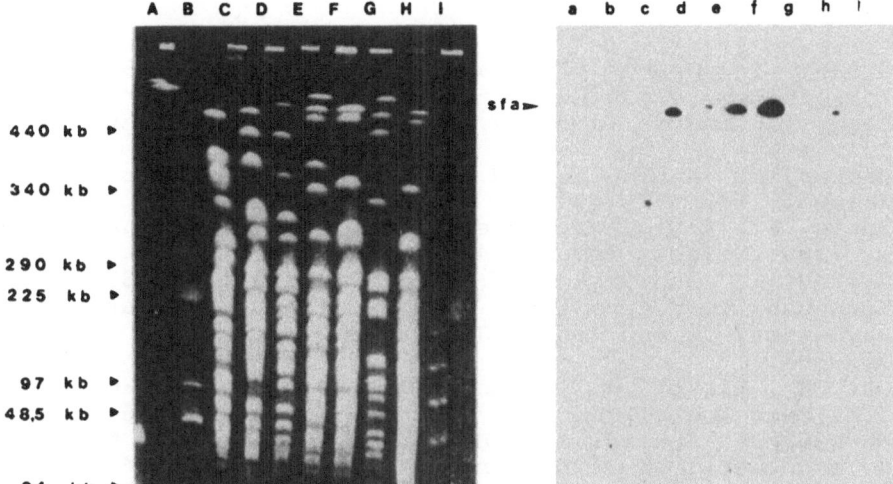

Fig. 3.   Pulse field analysis of *XbaI*-cleaved genomic DNAs
          (lanes A-I), and hybridization pattern of DNAs hybrid-
          ized with a *sfa* specific ³²P-labeled DNA probe (lanes
          a-i). The following strains are indicated: A159 (O2:K1,
          lanes B,b), RS226 (O18:K1, lanes C,c); IHE3034 (O18:K1,
          lanes D,d); 536-22 (O6:K15, lanes E,e); 536 (O6:K15;
          lanes F,f,H,h), HB101 (K-12, lanes G,g). As markers
          yeast chromosomes (lanes (A,a) and lambda-oligomeres
          (lanes I,i) are used.

like adhesins, capsules, O-antigens, iron uptake systems and
hemolysins contribute to pathogenicity of such extraintestinal *E.
coli* isolates. We have cloned and sequenced the genetic determinant responsible for the *E. coli* S fimbrial adhesin (*sfa*), an
attachment factor which is able to bind to Sialic acid-containing
receptor molecules. The *sfa* gene cluster codes for four different
fimbrial subunits, one of these, faS, a protein of 14 kd represents the S specific adhesin. With the help of the pRT733 vector
system we have constructed a site specific *sfa-lacZ* wild-type
fusion in the chromosome of the pathogenic *E. coli* strain 536.
Using this gene fusion it was shown that the wild-type expression
of *sfa* depends highly on environmental conditions like composition of the growth medium, temperature, oxygen osmolarity, and
growth phase. It was further demonstrated that S fimbriae contribute to nephropathogenicity of strain 536 in a rat pyelonephritis system. The expression of *sfa*, however, varied during
such an infection. A fa-negative mutant of strain 536, which was
isolated from the kidney of rats seven days after injection of
strains, was analyzed. Pulse-field gel electrophoresis studies
gave clear-cut evidence that chromosomal rearrangements and
deletions had occurred in this mutant, thereby influencing the
production of fa and other virulence factors.

ACKNOWLEDGEMENTS

     The authors wish to thank Balakrishna Pillay (Wurzburg) for
critical reading of the manuscript and Wilma Schmitt (Wurzburg)
for editorial assistance. The work was supported by the Deutsche
Forschungsgemeinschaft and the Fond der Chemischen Industrie.

REFERENCES

1. Orskov, I., Orskov, F., Jann,B., and Jann, K. 1976. Bacteriol. Rev. 41:667-710.
2. Bagg, A., and Neilands, J.B. 1987. Microbiol. Rev. 51: 509-534.
3. Hacker, J., and Hughes, C. 1985. Curr. Top. Microbiol. Immunol. 118:139-162.
4. Konig, B., Konig, W., Scheffer, J., Hacker, J., and Goebel, W. 1986. Infect. Immun. 54:886-892.
5. Orskov, I., and Orskov, F. 1983. Prog. Allergy 33, 80-105.
6. Korhonen, T. K., Vaisanen-Rhen, V., Rhen, M., Pere, A., Parkkinen, J. and Finne, J. 1984. J. Bacteriol. 159: 762-766.
7. Ott, M., Hacker, J., Schmoll, T., Jarchau, T., Korhonen, T.K., and Goebel, W. 1986. Infect. Immun. 54:646-653.
8. Hacker, J., Schmidt, G., Hughes, C., Knapp, S., Marget, M.,and Goebel, W. 1985. Infect. Immun. 47:434-440.
9. Schmoll, T., Hacker, J., and Goebel, W. 1987. FEMS Microbiol. Lett. 41:229-235.
10. Schmoll, T., Hoschutzky, H., Morschhauser, J., Lottspeich, F., Jann, K. and Hacker, J. 1989. Mol. Microbiol. 3:1735-1744.
11. Moch, T., Hoschutzky, H., Hacker, J., Kroke, K.-D., Jann, K. 1987. Proc. Natl. Acad. Sci. USA 84:3462-3466.
12. Klemm, P. 1985. Rev. Infect. Dis. 7:321-340.
13. Jacobs, A.A.C. 1987. Structure - function relationships of the K88 and K99 fibrillar adhesins of enterotoxigenic *Escherichia coli*. Diss. Univ. Amsterdam
14. Hacker, J. Jarchau, T., Knapp, S., Marre, R., Schmidt, G., Schmoll, T., and Goebel, W. 1986. In: Protein-Carbohydrate interactions - Molecular biology of microbial pathogenicity eds: D. Lark, S. Normark, H. Wolfwatz, B.E. Uhlin., Academic Press, London pp. 125-133.
15. Schmoll, T., Morschhauser, J., Ott, M., Luderiz, B., Van Die, I. and Hacker, J. 1990. Microb. Pathogen. 9: in press.
16. Miller, V.L., Taylor, R.K., and Mekalanos, J.J. 1987. Cell 48:271-279.
17. Schmoll, T., Ott, M., Oudega, B. and Hacker, J. 1989. J. Bacteriol. 172:5103-5111.
18. Marre, R., Hacker, J., Henkel, W., and Goebel, W. 1986. Infect. Immun. 54:761-767.
19. Knapp, S., Hacker, J., Jarchau, T., and Goebel, W. 1986. J. Bacteriol. 168:22-30.
20. Hacker, J., Schmoll, T., Ott, M., Marre, R., Hof, H., Jarchau, T., Knapp, S., Then, I., and Goebel, W. 1989. Stud. Iknfect. Dis. Res. eds:E. Kass and C. Svanborg-Eden, Univ. Chicago Press, Chicago, London pp.140-156.

ENTRY AND INTRACELLULAR BEHAVIOUR OF *SHIGELLA*

P. J. Sansonetti

Unite de Pathogenie, Microbienne Moleculaire and
Institut National de la Sante et de la Recherche
Medicale, U199, Institut Pasteur, 29 Rue du Docleur oux
75724 PARIS Cedex 15. FRANCE

INTRODUCTION

The pathogenesis of bacillary dysentery involves invasion of the human colonic mucosa (1,2). lnvasion encompasses
entry of the bacterium into epithelial cells, followed by
intracellular multiplication, intra and intercellular spread
and host cell killing (3). These bacteria then gain access to
the lamina propria of the colonic mucosa in   which a severe
inflammatory reaction causes abscesses and ulcerations. The
first step reflects the ability of the bacterium to induce
its own phagocytosis by cells which are non professional
phagocytes. It can be studied in vitro by infection of monolayers of mammalian cells such as Henle or HeLa cells (1,4,
5). The second step requires bacteria to   survive within
tissues and to elicit inflammation and tissue destruction. It
can be studied in assays such as the Sereny test (6), and
the rabbit ligated ileal loop (7). A modification of the
tissue culture assay allows quantitation of the microorgan-
isms` capacity to invade cells, multiply intracellularly and
spread to adjacent cells,   thus causing a cytopathic effect
visualized as clear plaques on a confluent monolayer (8).
Cell invasion by Shigella has recently been extensively
reviewed (3,9).

MECHANISMS OF ENTRY OF *S. FLEXNERI* INTO CELLS

Biology of the entry process

Studies using electron microscopy have shown that entry
into cells   occurs through a process which requires energy
production from the microorganism and host cells (4, 5).
Entry may therefore involve either receptor mediated endocy-
tosis which makes use of clathrin, the major coating component of coated pits and coated vesicles (10, 11), or phagocytosis, a property of polymorphonuclear cells and monocyte-
macrophages. Local polymerization of actin in microfilaments
(F- actin) and accumulation of myosin are the driving force
of this process (12, 13). Cytochalasin B and D which block
actin polymerization are potent inhibitors of phagocytosis

Microbial Surface Components and Toxins in Relation to Pathogenesis
Edited by E.Z. Ron and S. Rottem, Plenum Press, New York, 1991

101

(14). They inhibit entry of *Shigella* into cells (5). Accumulation of microfilaments at the site of entry of *S. flexneri* has even been directly demonstrated by fluorescence microscopy using NBD-phallacidin a dye which binds specifically to F-actin (15, 16). Myosin also accumulates at this site, as shown by indirect immunofluorescence using an antimyosin monoclonal antibody (16). Therefore, *S. flexneri* enters HeLa cells by directed phagocytosis. The transmembrane signals which elicit the phagocytic process are as yet unknown. Invasion of colonic epithelial cells also involves disruption of the brush border (17).

Genetic basis of the entry process

Plasmids of 220 kD are consistently found in *S. flexneri* (18) and in other *Shigella* species (19, 20). These plasmids are associated with the invasive phenotype. Sequences present on these plasmids encode the entry phenotype. In the case of *S. flexneri*, they have been cloned on a cosmid vector (21). Subsequent mutagenesis by Tn5 allowed definition of at least five independent loci spanning 20 kb (22). These date have been confirmed in *S. flexneri* and other *Shigella* species (23, 24). Hale et al. have shown that sera from monkeys and children convalescing from shigellosis recognize four plasmid-- encoded polypeptides (25). These polypeptides called ipaA (78 kD), ipaB (62 kD), ipaC (43 kD) and ipaD (38 kD), are also encoded by the recombinant cosmids which confer entry. Genetic organization of DNA sequences encoding ipa`s has been determined by analysis of a lambda-gt11 expression library (26) showing that three *ipa* genes (B, C and D) mapped to contiguous HindIII fragments; and by subcloning of the cosmid sequence along with Tn5 mutagenesis which showed a gene order of *ipa*B C D A (22). Sequencing of this region confirmed the gene order as the presence of two non-immunogenic polypeptides of 21 and 17 kD (27). The genes encoding these proteins are part of an operon.

The role of Ipa`s in the entry process is still unclear. Tn5 insertion which eliminate expression of *ipa*A do not affect entry whereas insertions which eliminate expression of *ipa* B, C and D render the strains non-invasive. Amino-acid sequence of *ipa*B and C deduced from their nucleotidic sequence, does not suggest any characteristic structural feature (27). No signal peptide can be detected and both proteins show a strong hydrophilic domain followed by a strong hydrophobic domain from N to C terminus. No homology has been detected with any other known DNA sequences. Nothing is known at present on the four other genetic loci involved in the entry precess of *S. flexneri* and how their gene products may interfere with Ipas. In *S. sonnei* subclones of two contiguous HindIII fragments express at least four polypeptides of 80, 47, 41 and 38 kD. Their genes map to the opposite end of the above mentioned sequence (24). Four loci may encode products allowing proper location or secretion of Ipas on the surface of the microorganism.

Regulation of the entry phenotype

There are at least two levels of regulation of expression of the entry phenotype. One locus, *virF*, is located on the virulence plasmid and acts *in trans* as a positive regulator (28). Another locus, *virR*, is located on the chromosome and acts *in trans* as a repressor (29). *VirF* which is located 40 kb apart from the entry genes, encodes a 30 kD protein which positively regulates *ipa* genes at the transcriptional level (28). *VirR* is located between *trp* and *galU* at 27 min. on the chromosome. It acts as a repressor of plasmid invasion genes in a temperature dependent manner since bacteria grown at 30°C are non invasive and do not express invasion-associated polypeptides whereas bacteria grown at 37°C are fully invasive and express such polypeptides. It is not yet known whether *virR* is a specific repressor of plasmid virulence functions or a gene which affects transcription of all plasmid genes. In any events, *Shigella* appears to use temperature (*virR*) as well as other unknown environmental signals (*virF*) to control expression of its invasion genes. Additional regulator functions will certainly be demonstrated in a near future.

MOLECULAR MECHANISMS OF INTRACELLULAR MULTIPLICATION OF *SHIGELLA*

Plasmid genes are also needed for efficient intracellular multiplication (30). This capacity to multiply rapidly after entry into cells is a characteristic of *Shigella* as compared to other enteroinvasive microorganisms such as *Yersinia* or *Salmonella* (31, 32). A sequential electron microscopy study performed on HeLa cells infected by an invasive isolate of *S. flexneri* demonstrated lysis of the phagocytic membrane occurring early after entry into cells (30). The plasmid-mediated contact hemolytic activity expressed by *Shigella* is a likely candidate for lysis of the phagosome (30, 33). Any Tn5 mutant which eliminates the contact-hemolytic phenotype also eliminates the entry phenotype and vice-versa (Baudry et al., unpublished date). This suggests that the surface construction which elicits entry into cells also accounts for lysis of the phagocytic vacuole due to rapid intraphagosomal acidification. Drop in pH from 7 to 5.5 has been shown to enhance hemolytic activity by at least 100 fold (33). Conflicting data have recently been published according to which lysis of the phagocytic vacuole would be caused by the product of *kcpA*, a chromosomal locus located at 13 min on the chromosome (34).

GENETIC AND MOLECULAR BASIS OF INTRACELLULAR SPREAD AND INFECTION OF ADJACENT CELLS

The capacity to spread intracellularly and infect adjacent cells is of critical importance for the outcome of the infection during shigellosis. This capacity is reflected in the plaque assay in which invading microorganisms express their capacity to spread from one cell to another and ultimately cause a cytopathic effect through host-cell killing (8). We have isolated a TnphoA mutant of *S. flexneri* (SC557) which invades HeLa cells but does not from plaques (35). SC557 localizes within the cytoplasm of infected cells without evidence of intracellular spread and infection of adjacent cells. It has retained its capacity to lyse the membrane bound phagocytic vacuole but has lost its property to utilize polymerized actid (F-actin) to spread intracellularly and infect adjacent cells. A similar localization of the microcolony is observed when cells recently infected by the wild type microorganism are treated with cytocholasin D, which blocks polymerization of globular actin (G-actin) to F-actin. These date confirm that direct interaction of the invading microorganism with cell microfilaments is necessary to ensure intracellular spread and infection of adjacent cells. A plasmid gene, icsA (intra-intercellular spread) has been identified in SC557 which encodes a 120 kD outer membrane protein. This protein is recognized by sera from monkeys convalescing from shigellosis. Its role in the spreading process is currently under study.

These results are consistent with a previous publication in which a plasmid locus, *vir*G is necessary for permanent reinfection of adjacent cells (36). The *kcp*A locus necessary to elicit a kerato-conjunctivitis in guinea pigs (37), might be a regulatory function which positively controls expression of *ics*A (*vir*G) (38).

EARLY KILLING OF HOST CELLS BY *SHIGELLA*

Cytotoxicity of *Shigella* can also be studied in vitro in epithelial cells. However, use of the macrophage J774 allowed study of the intracellular fate of both an invasive and a non-invasive derivative of *S. flexneri*. The virulence plasmid appeared to mediate efficient and rapid killing of host-cells. Damage to macrophages correlated with the capacity of invasive bacteria to rapidly lyse the membrane of the phagocytic vacuole (39). Mitochondria may be a target for this activity.

Metabolic events mediating early killing involve a rapid drop in intracellular concentrations of ATP and an increase in pyruvate concentrations (40). Shiga toxin, a potent cytotoxin produced at a high level by *S. dysenteriae*1, does not play a significant role in early killing of infected cells. This is confirmed by testing a *Tox*- mutant of this species which does not produce *Shiga* toxin (41) . This mutant infected cells as efficiently as the wild type strain, indicating that the process of invasion is sufficient to account for early cell killing. Secretion of *Shiga* toxin within infected tissues causes severe alterations of the capillaries in the colonic lamina propria.

CONCLUSIONS AND PERSPECTIVES

Molecular and cellular biology are providing insights into the process of epithelial-cell invasion by *Shigella*. This microorganism has evolved a several-stage strategy encompassing entry through directed phagocytosis, lysis of the membrane bound phagocytic vacuole allowing access to the cytosol in which it grows rapidly. It subsequently spreads

intra and intercellularly, interacting with host-cell microfilaments. It will eventually kill host cells by a process
involving blocking of respiration but not production of *Shiga*
- toxin. This process makes the pathogenesis of shigellosis
quite different from that of salmonellosis. The way these
two species infect the intestinal epithelium has a strong
influence on the outcome since *Shigella* uses the epithelium
as its site of multiplication, whereas *Salmonella* uses it as
a route to gain access to the lamina propria and mesenteric
lymph nodes through transcytosis (42). In addition to understanding the complex pathogenic process of shigellosis,
these informations may be of primary importance to design
live oral vaccine strains to protect against shigellosis.

REFERENCES

1. LaBrec, E. H., Schneider, H., Magnani, T.J., Formal, S.B. 1964. J. Bacteriol. 88:1503-1518.
2. Takeuchi, A,. Formal, S.B,. Sprinz, H. 1968. Am. J. Pathol. 52:503-519.
3. Maurelli, A. T., Sansonetti, P. J. 1988. Ann. Rev. Microbiol. 42:127-150.
4. Hale, T. L., Bonventre, P. F. 1979. Infect. Immun. 24: 879-886.
5. Hale, T, L., Morris, R. E. Bonventre, P. F. 1979. Infect Immun. 24: 887-894.
6. Sereny, B. 1957. Acta Microbiol. Acad. Sci. Hung. 4:367-376.
7. Formal, S. B., Kundel, D., Schneider, H., Kunev, N., Sprinz, H. 1961. Br. J. Exp. Pathol. 42:504-510.
8. Oaks, E. V., Wingfield, M. E., Formal, S. B. 1985. Infect. Immun. 53:57-63.
9. Hale T. L. Formal, S. B. 1986. Microb. Pathogen. 1: 511-518.
10. Pearse, B. M. F. Clathrin. 1976. Proc. Natl. Acad. USA. 73:1255-1259.
11. Goldstein, J. L., Anderson, R. G. W., Brown, M. 1979. Nature (London). 279: 679-685.
12. Stendahl, O. I., Hartwin, J. H., Brotschi, E. A., Stossel, T. P. 1980. J. Cell. Biol. 84:215-224.
13. Sheterline, P., Richard, J. E., Richards, J. C. 1984. Eur. J. Cell. Biol. 34: 80-87.
14. Tanenbaum, S. W. (ed.). Cytocholasins: biochemical and cell biological aspects. 1978. North-Holland Publishing Co., Amsterdam.
15. Barak, J. S., Yoam, R. R., Nothnagel, E. A., Webb, W. W. 1980. Proc. Natl. Acad. Sci USA. 77: 980-984.
16. Clerc, P., Sansonetti, P. J. 1987. Infect. Immun. 55: 2681-2688.
17. Polotsky, Y. E., Snigevskaya, E. S., Dragunskaya, E. M. Bull. Exp. Med. (Moscow). 1974. 77:110-114.
18. Sansonetti, P. J., Kopecko, D. J., Formal, S. B. 1982. Infect. Immun. 35:852-860.
19. Sansonetti, P. J., Kopecko, D. J., Formal, S. B. 1981. Infect. Immun. 34:75-83.
20. Sansonetti, P. J., d`Hauteville, H., Ecobichon, C., Pourcel, C. 1983. Ann. Microbiol. (paris). 134A:295-318.
21. Maurelli, A. T., Baudry, B. d'Hauteville, H., Hale, T. L. Sansonetti, P. J. 1985. Infect. Immun. 49:164-171.

22. Baudry, B., Maurelli, A. T., Clerc, P., Sadoff, J. C., Sansonetti, P. J. 1987. J. Gen. Microbiol. 133:3409-3413.

23. Sasakawa, C., Kamata, K., Sakai, T., Makino, S., Yamada, M., Okada, N., Yoshikawa, M. 1988. J. Bacteriol. 170: 2480-2484.

24. Watanabe, H., Nakamura, A. 1986. Infect. Immun. 53:352-358.

25. Hale, T. L., Oaks, E. V., Formal, S. B. 1985. Infect. Immun. 50:620-629.

26. Buysse, J. M., Stover, C. K., Oaks, E. V., Venkatesan, M., Kopecko. D. J. 1987. J. Bacteriol. 169:2561-2569.

27. Baudry, B., Kaczorek, M., Sansonetti, P. J. 1988. Microb. Pathogen. 4:345-357.

28. Sakai, T., Sasakawa, C., Yoshikawa, M. 1988. Mol. Microbiol. 2:589-597.

29. Maurelli, A. T., Sansonetti, P. J. 1988. Proc. Natl. Acad. Sci. USA. 85:2820-2824.

30. Sansonetti, P. J., Ryter, A., Clerc, P., Maurelli, A. T., Mounier, J. 1986. Infect. Immun. 55:521-527.

31. Small, P. L. C. Isberg, R. R., Falkow, S. 1987. Infect. Immun. 55:1674-1679.

32. Finlay, B. B., Falkow, S. 1988. UCLA Symp. Mol. Cell. Biol. 64: 227-24.

33. Clerc, P., Baudry, B., Sansonetti, P. J. 1986. Ann. Inst. Pasteur Microbiol. 134A:267-278.

34. Yamada, M., Sasakawa, C., Okada, N., Makino, S.I., Yoshikawa, M. 1989. Mol. Micro. 3:207-213.

35. Bernardini, M.L., Mounier, J., d`Hauteville, H., Coquis-Rondon, M., Sansonetti, P. J. 1989. Proc. Nat. Acad. Sci. (in press).

36. Makino, S., Sasakawa, C., Kawata, K., Kurata, T., Yoshikawa, M. Cell. 46:551-555.

37. Formal, S. B., Gemski, P. Jr., Baron, L. S., LaBrec, E. H. A. 1971. Infec. Immun. 3:73-79.

38. Pal., Newland, J. W., Tall, B. D., Formal, S.B., Hale, T. L. 1989. Infect. Immun. 57:477-486.

39. Clerc, P., Ryter, A., Mounier, J., Sansonetti, P. J. 1987. Infect. Immun. 48:124-129.

40. Sansonetti, P.J., Mounevir, J. 1987. Microb. Pathogen. 3:53-61.

41. Fontaine, A., Arondel, J., Sansonetti, P. J. 1988. Infect. Immun. 56:3099-3109.

42. Finlay, B.B., Gunbiner, B., Falkow, S. 1988. J. Cells. Biol. 107: 221-230.

THE *SALMONELLA TYPHIMURIUM* VIRULENCE PLASMID

Mikael Rhen, Suvi Taria, Soila Sukupolvi,
Marita Virtanen and P. Helena Makela

Molecular Biology Unit, National Public
Health Intitute, Mannerheimintie 166
00300 Helsinki, Finland

INTRODUCTION

Salmonellosis is a good example of an infectious disease in which intracellular growth is an essential step; *Salmonellae* undergo a major multiplication step within host tissue macrophages. Many different serovars of *Salmonella* cause extraintestinal infections in man and mouse, yet the course of the infection is similar, if not identical.

The infecting organism is acquired from ingested food or water whereafter the bacteria that managed to get into the small intestine adhere to the intestinal epithelium (1-3). They then penetrate the epithelium, preferentially at the Payer's patches. Bacterial multiplication takes place in these and local lymph nodes resulting in a primary bacteremia (2-4). From the blood bacteria are efficiently taken up by macrophages of the liver and spleen. In contrast to many other bacteria virulent *Salmonellae* not only survive within these macrophages but even undergo a net multiplication resulting in a final fulminant septicemia when the bacterial load becomes large enough (2,4,5).

The infection pathogenesis of salmonellosis is thus fairly well known on a phenomenological level. *S. typhimurium* that causes salmonellosis in the mouse is also well characterized both biochemically and genetically. Experimental mouse salmonellosis is therefore an excel lent experimental system for studying the molecular biology of intracellular parasitism.

Serovars of *Salmonella* that cause salmonellosis in mice harbour a large, non-conjugative plasmid (5-11). Virulent *Salmonella* strains can be made avirulent by curing them of this plasmid. The infection with such cured strains progress normally through the early steps, until the point of multiplication within host macrophages (4). The cured bacteria are apparently internalized normally, but their subsequent net multiplication is prevented. The large plasmid of these bacteria therefore seems to encode functions essential for intracellular growth.

Microbial Surface Components and Toxins in Relation to Pathogenesis
Edited by E.Z. Ron and S. Rottem, Plenum Press, New York, 1991

STRUCTURAL AND PHYSIOLOGICAL PROPERTIES ASSOCIATED WITH THE VIRULENCE PLASMID

Wild-type *S. typhimurium* strains and their cured derivatives behave almost identically under laboratory conditions. In fact it has been difficult to conclusively demonstrate any plasmid-encoded functions or proteins besides the quite descriptive concept of virulence. The first, and for a long time the only phenotype associated with this plasmid was fertility inhibition (fi) which sometimes hampered genetic experiments in *Salmonella* (12,13). Therefore the plasmid was long called "cryptic".

One of the first potential virulence characteristics associated with the virulence plasmid was serum resistance. Strains of *Salmonella* harbouring the plasmid tolerate serum better that their cured derivatives (7,14,15). This effect has been correlated with complement activity in the sera; the serum tolerance should be better described as complement resistance (15). The location where the plasmid is supposed to act is not in blood or serum but in the intracellular compartments of macrophages. Macrophages do synthesize complement components (16), which might also act in the intracellular vesicles. However, complement resistance could be an indication of a more general physicochemical stability of the outer membrane. This might be required to resist defensins - the cationic defence proteins of vertebrates contained in macrophagal phagolysosomes (17).

Hackett and coworkers were the first to analyze the virulence plasmid-mediated serum resistance of *S. typhimurium* on gene level (14). They isolated by molecular cloning a 2.1-kb plasmid fragment encoding an 11-kDa outer membrane protein that conferred resistance to human serum in both *S. typhimurium* and *E. coli*. In addition, the plasmid has been reported to contain a 66 base pair regulatory element essential for resistance to guinea pig serum (18). The interpretation of these results has been challenged by Gulig and Curtiss; human and rabbit sera did not differentiate their wild-type *S. typhimurium* and its cured derivative (6).

Our own investigations indicate that the virulence plasmid of *S. typhimurium* line TML increases resistance of the bacteria towards guinea pig serum (19). However, smooth, cured *Salmonella* strains still exhibit a higher serum resistance that rough *E. coli* K-12 strains. We have cloned and characterized one gene of the *S. typhimurium* virulence plasmid that increases resistance towards guinea pig serum both in *S. typhimurium* and *E. coli* (19,20). The gene encodes a 28-kDa outer membrane protein and maps at the plasmid coordinate 67 kb (Figure 1).

Immunochemical and hybridization analyses indicated that the corresponding gene encoded a TraT-like protein. This conclusion was verified by nucleotide sequence analysis of the corresponding gene. It encoded a 243 amino acid protein including a signal sequence of 20 residues (21). The protein showed 91.4% homology with the TraT proteins of factor F and the antibiotic resistance plasmid R100 on amino acid sequence level (21). This is not surprising since *traT* genes of antibiotic resistance plasmid are known to increase serum

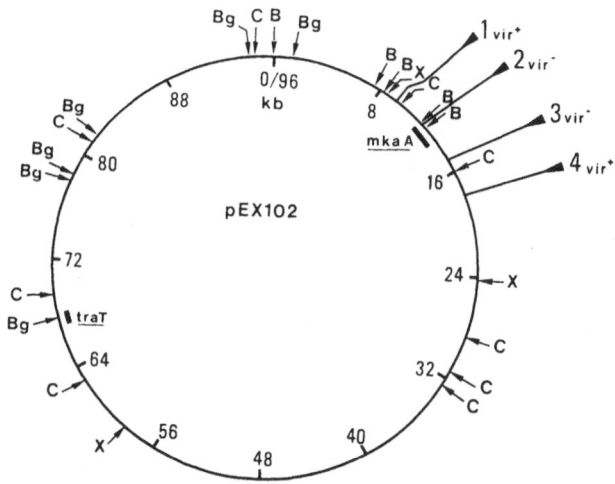

Fig. 1. Restriction endonuclease map of the *S. typhimurium* virulence plasmid pEX102. Restriction endonuclease abbreviations: B=*Bam*HI. Bg=*Bgl*II, C=*Cla*I and X=*Xho*I. The numbers inside the ring are map coordinates in kb. The positions of two virulence--abolishing Tn5 insertions (*zzx-2556* and *zzx-2558*) are shown as arrows 2 and 3, respectively.

resistance; they have in fact been proposed to contribute to virulence in *E. coli* causing extraintestestinal infections (22,23). Insertional inactivation of the *traT* gene on the *S. typhimurium* virulence plasmid resulted in decreased serum resistance but not decreased virulence (19).

In 1984 Sukupolvi and coworkers (24) described a novel type of outer membrane permeability in *S. typhimurium*. The mutant strain was characterized by increased sensitivity towards hydrophobic antibiotics but expressed the parental type lipopolysaccharide (LPS) and outer membrane protein profile. The mutation was mapped into the *traT* gene of the virulence plasmid (20,25).

A concomitant loss of smooth type LPS and the virulence plasmid has been reported by Hackett and coworkers (14). This association is, however, probably spurious since other cured derivatives of wild-type *S. typhimurium* still possess complete O antigens and LPS core regions (6,18) (Figure 2). Other reported potential virulence properties encoded by the plasmid include adherence to and invasion into HeLa cells (26,27). Unfortunately, these observations have not been confirmed (3) or studied on gene level.

INTRACELLULAR GROWTH AND THE VIRULENCE PLASMID

Salmonella bacteria given intravenously are rapidly taken up by the macrophages of the liver and spleen of mice (11). Plasmid-containing, virulent strains then multiply during the subsequent days and kill the animal when the number of the bacteria are approximately 10^8 per mouse; with wild-type bacteria and mice this takes about five days after an inoculum of 10^5 bacteria. In identical experiments with

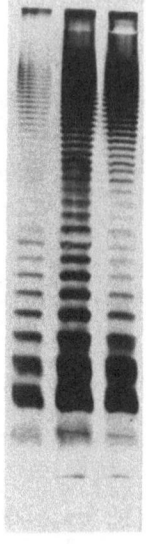

Fig. 2.   Silver-stained SDS-polyacrylamide gels of LPS prep-
            arations from *S. typhimurium* strains. All strains
            exhibit smooth type laddern pattern of LPS and
            sensitivity to phage p22.

cured *Salmonella* strains the numbers of bacteria in the liver
and spleen do not increase (Figure 3) and the animals survive
challenge doses of 10^6. The virulence can be restored by
reintroducing the plasmid into the cured strain (Figure 3).

        In an attempt to more precisely mimic the natural infec-
tion Hackett and coworkers fed mice with *S. typhimurium* and
with its cured derivative (4). Soon after challenge both
strains recovered in equal number from the liver and spleen,
and Peyer's patches. However, only the plasmid-containing
derivative produced net growth in the liver and spleen.
Together these experiments indicate that the plasmid encoded
functions needed for growth within macrophages of the liver
and spleen, whereas genes for other virulence functions, such
as epithelial cell penetration may be chromosomally encoded
(3).

        Both deletion and insertion mutagenesis has been em-
ployed to map plasmid regions responsible for intracellular
growth. Michiels and coworkers (27) used a set of deletion
derivatives of the virulence plasmid to map functions associ-

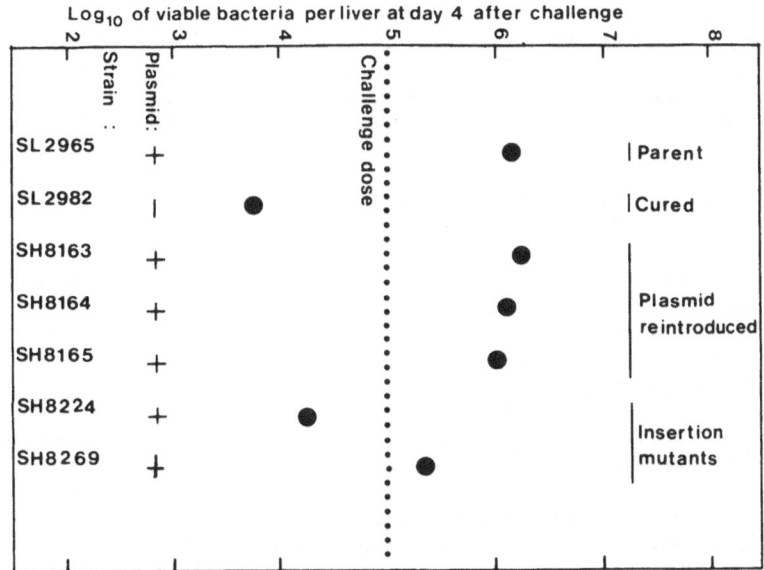

Fig. 3. Growth of *Salmonella* in the liver of (CBA x C5781-/6)F1 mice. Challenges (10^5 bacteria/mouse) were give intravenously and viable bacteria of the livers enumerated four days after this challenge. Strains SH8224 and SH8269 harbour the Tn5 insertions 2 and 3, respectively, in their virulence plasmids.

ated with virulence and replication. Furthermore, a restriction map of the entire c. 100-kb virulence plasmid was constructed from the data of the various deletion derivatives. The important finding was that the virulence genes were not uniformly distributed over the plasmid but rather clustered within a 32-kb region. Additionally, the plasmid contained at least two origins of replication.

Transposon insertion mutagenesis of the *S. typhimurium* virulence plasmid had identified two virulence-associated regions (28,29). These regions are closely linked and, within the resolution of the mutagenesis strategy, possibly represent distinct genes of a larger more or less continuous virulence region. This region encompasses less than 10 kb (map co-ordinates 9-16 in Figure 1) and falls entirely within the region defined by Michiels and coworkers (27).

Information about specific genes within the defined virulence region is so far only scant. Gulig and Curtis have identified one 28-kDa protein encoded by this region that appears essential for the virulence (28). We have studied the effect of two transposon insertions located in the neighbouring virulence region (Figure 1) Both these insertions decrease the intracellular growth rate of the bacteria (29,30) (Figure 3). The Tn5 insertion 2 inhibits the expression of a 70-kDa protein when studied in minicells containing the gene cloned in pBR325. The corresponding gene has been sequenced; it en codes a 591 amino acid protein. Insertion 2 is located between the codon triplets for amino acid residues 374 and 375 (30). The second transposon (3 in Figure 1) is some 3.5 kb away from 2; it inhibits the expression of a 30 kDa protein from appropriate recombinant plasmid. The precise defect caused by insertion 3 remains to be determined.

RELATIONSHIP OF VIRULENCE PLASMID FROM VARIOUS SEROVARS OF *SALMONELLA*

The presence of a virulence plasmid is by no means restricted to *S. typhimurium*. Functionally analogous plasmids have been described in e.g. the serovars *S. dublin* (7,10,31). *S. pullorum* (32), *S. enteritidis* (7,9,,11) and *S. cholerae-suis* (33). That is to say, in all these instances the plasmid seems to be required for intracellular growth. The virulence plasmid of *S. dublin* and *S. enteritidis* have also been reported to mediate serum resistance (7,34).

Several investigations indicate, that the various virulence plasmids contain homologous DNA regions although the molecular sizes in the different serovars may differ (7,8, 31,35-37). The DNA region encompassing the virulence region in the plasmid of *S. typhimurium* (the region encompassing co-ordinates 5 to 15 in Figure 1) detected upon DNA hybridization a homologous region in all the serovars listed above (37). Similarly, the virulence region of the plasmid of *S. dublin* appears, as judged by restriction maps, homologous to that of *S. typhimurium* and is likewise present in a number of plasmid-containing serovars (31,36,37). The DNA region outside the virulence region is present to a large extent in the various virulence plasmids. In fact, we have been able to restore mouse virulence in a cured *S. enteritidis* strain by introducing the *S. typhimurium* virulence plasmid pEX102 (11). The large plasmid of *Salmonellae* required for intracellular growth have evidently evolved from a common ancestor and encode similar functions.

ACKNOWLEDGEMENTS

This work was supported by the Sigrid Juselius Foundation (MR, ST, MV, PHM) and the Finnish Cultural Foundation ss.

REFERENCES

1.   Takeuchi, A. 1967. Am.J.Pathol. 50:109-136.
2.   Heffernan E.J., Fierer, J., Chikami, G. and Guiney, G. 1987. J. Infect. Dis. 155:1254-1259.
3.   Finlay, B,B., Stambach, M.N., Francis, C.L., Stocker, B.A.D., Chatfield, S., Dougan, G. and Falkow, S. 1988. Molec. Microbiol. 2:757-766.
4.   Hackett, J., Kotlarsky, I., Mathan, V., Francki, K. and rowley, D. 1986. J. Infect. Dis. 153:1119-1125.
5.   Fields, P.I., Swanson, R.V., Haidairs, C.G. and Heffron, F. 1986. Proc. Natl. Acad. Sci. USA. 83:5189-5193.
6.   Gulig, P.A. and Curtiss, R III. 1987. Infect. Immun. 55: 2891- 2901.
7.   Helmuth, R., Stephen, R., Bunge, C., Hoog, B.,

Steinbeck, A. and Bulling, E. 1985. Infect. Immun. 48: 175-182.

8. Pardon, P., Popoff, M.Y., Coynault, C., Marly, J. and Miras, J. 1986. Ann. Microbiol. Inst. Pasteur. 137B: 47-60.

9. Nakamure, M., Sato, S., Ohya, T., Suzuki, S. and Ikeda, S. 1985. Infect. Immun. 47:832-833.

10. Terakado, N., Sekizaki, T., Hashimoto, K. and Naitoh, S. 1983. Infect. Immun. 41:443-444.

11. Hovi, M., Sukupolvi, S., Edwards, M.F. and Rhen, M. 1988. Microb. Pathogen. 4:385-391.

12. Anderson, E.S. and Smith, H.R. 1972. Mol. Gen. Genet. 118:79- 84.

13. Smith, H.R., Humphreys, G.O., Grindley, N.D.F. and Anderson, E.S. 1973. Mol. Gen. Genet. 126:143-151.

14. Hackett, J., Wyk, P., Reeves, P. and Mathan, V. 1987. J. Infect. Dis. 155:540-549.

15. Vandenbosch, J.L., Rabert, D.K. and Jones, G.W. 1985. Infect. Immun. 55:2645-2652.

16. Hetland, G. and Eskeland, T. 1986. Scand. J. Immunol. 23:301- 308.

17. Viljanen, P., Koski, P. and Vaara, M. 1986. Infect. Immun. 56:2324-2329.

18. Vandenbosch, J.L., Rabert, D.K., Kurlandsky, D.R. and Jones, G.W. 1989. Infect. Immun. 57:850-857.

19. Rhen, M. and Sukupolvi, S. 1989. Microb. Pathog. 5: 275-285.

20. Rhen, M., O`Connor, C.D. and Sukupolvi, S. 1988. FEMS Microbiol. Lett. 52:145-154.

21. Sukupolvi, S., Rhen, M., Vuorio, R. and O`Connor, C.D. 1989. Molec. Microbiol. submitted.

22. Moll, A., Manning, P.A. and Timmis, K.N. 1980. Infect. Immun. 28:359-367.

23. Montenegro, M.A., Bitter-Suermann, D., Timmis, J.K., Aguero, M.E., Cabello, F.C., Sanyol, F.C. and Timmis, K.N. 1985. J. Gen. Microbiol. 131:1511-1521.

24. Sukupolvi, S., Vaara, M., Helander, I.M., Viljanen. P. and Makela, P.H. 1984. J. Bacteriol. 159, 704-712.

25. Sukupolvi, S., O`Connor, C.D. and Edwards, M.F. 1986. J. Gen. Microbiol. 132, 2079-2085.

26. Jones, G.W., Rabert, D.K., Svinarice, D.M. and Whitfield, H.J. 1982. Infect. Immun. 38, 476-486.

27. Michiels, T., Popoff, M.Y., Durviaux, S., Coynault, C. and Cor nelis, G. 1987. Microb. Pathog. 3, 109-116.

28. Gulig, P.A. and Curtiss, R III. 1988. Infect. Immun. 56, 3262- 3271.

29. Rhen, M., Virtanen, M. and Makela, P.H. 1989. Microb. Pathog. 6, 153-158.

30. Taira, S. and Rhen, M. 1989. Microb. Pathog. in press.

31. Beninger, P.R., Chikami, G., Tanabe, K., Roudier, C., Fierer, S. and Guiney, D.G. 1988. J. Clin. Invest. 81, 1341-1347.

32. Barrow, P.A. and Lovell, M.A. 1988. J. Gen. Microbiol. 1234, 2307-2316.

33. Kawara, K., Haraguchi, Y., Tsuchimoto, M., Terakado, N. and Dan bara, H. 1988. Microb. Pathog. 4, 155-163.

34. Manning, J.E., Baird, G.D. and Jones, P.W. 1986. J. Med. Microbiol. 21, 239-243.

35. Baird, D.G., Manning, J.E. and Jones, P.W. 1985. J. Gen. Microbiol. 131, 1815-1823.

36. Williamson, C.M., Baird, D.G. and Manning, J.E. 1988. J. Gen. Microbiol. 134, 975-982.
37. Korpela, K., Ranki, M., Sukupolvi, S., Makela, P.H. and Rhen, M. 1989. FEMS Microbiol. Lett. 58, 49-58.
38. Helander, I. 1985. In: Korhonen, T.K., Dawes, E.A. and Makela, P.H. eds. Enterobacterial Surface Antigens. Methods for Molecular Characterization. Elsevier Science Publishers Biomedical Division.

REGULATION AND FUNCTION OF PLASMID ENCODED

VIRULENCE DETERMINANTS OF *YERSINIA*

H. Wolf-Watz, A. Forsberg, R. Rosqvist, I. Bölin, K. Erickson, L. Norlander, M. Rimpilainen, and T. Bergman

Department of Microbiology, University of Umeå
S-901 87 Umeå, Sweden and Department of Cell and
Microbiology, FOA4,S-901 82 Umeå, Sweden

INTRODUCTION

The three virulent members of the genus *Yersinia* harbour related virulence plasmids with a molecular weight of about 60-70 kb (1). When exponential phase cultures of these organisms growing in a Ca^{2+} free medium are shifted from 26°C to 37°C, growth ceases over a period of about 2 generations (2,3,4). If however 2.5 mM Ca^{2+} is present in the medium growth continues normally. These bacteria are referred to as being Ca^{2+} dependent (CD). Plasmid cured strains, however, do not show this dependency on Ca^{2+} and are thus, Ca^{2+} independent (CI). Such bacteria are always avirulent. By transposon insertion mutagenesis a 20 kb region of the virulence plasmid has been identified which is involved in this low calcium response (*lcr*). Such CI-mutants do not require Ca^{2+} for prolonged growth at 37°C and they are not virulent. Although the plasmids of *Y. enterocolitica, Y. pestis* and *Y. pseudotuberculosis* have been subjected to rearrangements, the Ca^{2+} region, however, of the different plasmids is conserved (Fig.1) (1).

EXPRESSION OF VIRULENCE PLASMID ENCODED VIRULENCE DETERMINANTS

When the plasmid containing strain of *Y. pseudotuberculosis*; YPIII(pIB1) is grown in a medium free of Ca^{2+} and shifted from 26°C to 37°C, the bacteria induce a number of temperature inducible proteins (Yops) (Fig. 2). These proteins can be recovered in a secreted form as well as in a membrane bound form (5). If Ca^{2+} is added to the culture the rate of synthesis of the Yops is greatly reduced and these proteins can only be detected in minute amounts in the membrane fraction by Western blotting (5,6). Thus, expression of the Yops is regulated by Ca^{2+} and temperature. All three species are able to express the Yops, showing the same regulatory pattern. In strains of *Y. pestis* however, the isolation of the Yops both from the culture supernatant and

Microbial Surface Components and Toxins in Relation to Pathogenesis
Edited by E.Z. Ron and S. Rottem, Plenum Press, New York, 1991

115

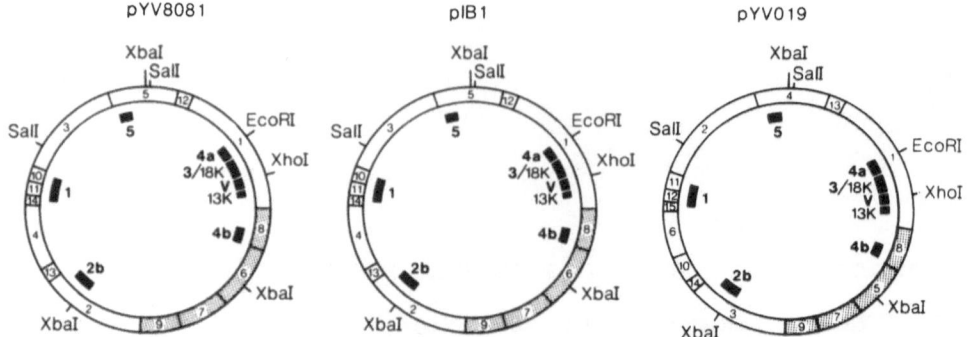

Fig. 1. Circular BamH1 restriction maps of plasmids pYV8081 of *Y. enterocolitica*, pYV019 of *Y. pestis* and pIB1 of *Y. pseudotuberculosis*. The stippled area indicates the Ca²⁺ region. Black bars show the position of identified gene products.

from the membrane fraction is hampered by the fact that these strains, in addition, express an extracellular protease which specifically degrades the Yops but not the integral outer membrane proteins (7). This enzyme (plasminogen-activator/-coagulase) is encoded by an additional 12 kb plasmid pCP1. Consequently, the Yops can easily be detected when pCP1 cured strains of *Y. pestis* are analyzed (8). Since the Yops are more sensitive to proteolysis compared to integral outer-membrane proteins it is possible that the Yops are not integral outer-membrane proteins but rather loosely attached to the bacterial surface. Upon comparison of the Yops from the three species it is found that they are very similar, showing antigenic relatedness and only minor differences with respect to their respective molecular weights (Table 1) (6). There-fore, it is highly likely that the Yop proteins exhibit the same virulence functions in all three species. Antibodies directed against the Yops can be detected in convalescent sera obtained from patients recovering from Yersinosis as well as bubonic plaque, demonstrating that the Yops are expressed during infection, which support the idea that at least some of these proteins are essential virulence determi-nants (9).

CLONING AND MAPPING THE *YOP* GENES

A *BamHI* gene-bank of pIB1 of *Y. pesudotuberculosis* was made. This bank was screened, in *E. coli* minicells for the ability of individual clones to express the Yops, by immuno-precipitation using a rabbit-anti-yop antiserum. A number of positive clones could be isolated and further analysis of these revealed that *Yop2b, Yop3, Yop4a, Yop4b, Yop5* and the V-antigen had been cloned (5). The corresponding genes were mapped on pIB1 and found to be scattered around the plasmid (Fig.1). At least the *yopH* and the *yopE* genes were found to be monocistronic while the structural genes of the Yop3, Yop4 and V-antigen proteins were closely linked (Fig.1). In accordance with earlier results (10), these latter genes were also found to be linked to two other genes encoding proteins

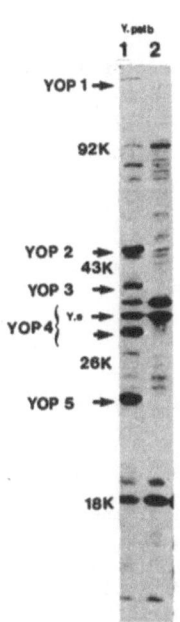

Fig. 2.   Outer-membrane protein profile of strains YPIII-
          (pIB1) and YPIII. The strains were grown in rich
          medium lacking Ca²⁺ at 37°C and the OM was prepared
          and dissolved in SDS-sample buffer. The samples
          were thereafter subjected to SDS-PAGE and stained
          with Coomassie blue. Lane 1. YPIII(pIB1), Lane 2.
          YPIII. Yop2 denotes the position of Yop2a and Yop2b
          and Yop4 denotes the position of Yop4a and Yop4b
          respectively. Y.e. shows the position of Yop4a of
          Y.enterocolitica. For molecular weighs see table 1.

having a molecular weight of 13 kdal (*lcrG*) and 18 kdal
(*lcrH*) respectively. Earlier studies had also shown that
these two genes in addition to the structural gene, *lcrV*, of
the V-antigen constituted a part of a common operon- *lcrGVH*
(10). To investigate whether the *yopC* and *yopD* genes also
were part of this operon we constructed a number specific
insertions mutants of *Y. pseudotuberculosis*; YPIII(pIB13)
lcrH, YPIII(pIB14) *yopC* and YPIII(pIB15) *yopD*. These mutants
showed all a temperature-sensitive (TS) phenotype and the
ability to form colonies on Ca²⁺ containing agar plates at
37°C was greatly reduced compared to the isogenic wild type
strain YPIII(pIB103). When the ability to secrete the Yops to
the culture medium of the different mutants was analyzed it
was found that the *lcrH* and the *yopC* mutants did neither
express Yop3 nor Yop4a while the *yopD* mutant did not express
Yop4a but was able to express Yop3. These results suggest
that the *yopC* and *yopD* genes in addition to the *lcrGVH* genes
constitute a operon having the gene order; *lcrG*, *lcrV*, *lcrH*,
yopC and *yopD*, respectively. The different genes were also
mapped on pYV8081 and pYV019 of *Y. enterocolitica* and *Y.
pestis* respectively by the use of Southern blotting (6). The
Y. pestis plasmid pYV019 was found to be homologous to pIB1
while pYV8081 showed differences upon comparison with pIB1
and pYV019. It is apparent that the plasmids have been subjected to rearrangements while all coding units have been.

TABLE 1. Sizes of the immunologically related YOPs and the V antigen of Yersinia spp.[a]

| Protein | Gene | Sizes (kDa) of proteins in the following strains: | | |
		Y. pseudotuberculosis YPIII(pIB1)	Y. pestis EV76	Y. enterocolitica 8081
YOP2a	yopM	45	44	44
YOP2b	yopH	45	45	46
YOP3	yopC	41-42[b]	41-42[b]	40-41[b]
V antigen	lcrV	38	38	38
YOP4a	yopD	34	34	36
YOP4b	lcrE	34	34	34
YOP5	yopE	26	26	25.5

[a] Sizes were determined by SDS-PAGE.

[b] The range in molecular sizes is due to the presence of two different formes of YOP3.

maintained. The largest stretch of uninterrupted homology between the plasmids contained the Ca^{2+} region (Fig.1).

REGULATION OF THE *YOP* GENES

The *yopE* (Yop5) and *yopH* (Yop2b) genes of *Y. pseudotuberculosis* strain YPIII(pIB1) were sequenced and both genes were found to be monocistronic (11,12). The transcription of these genes was studied by Northern blotting using bacteria incubated at 26°C or at 37°C in a medium containing or not containing Ca^{2+}. It was apparent that the expression of Yop2b and Yop5 was regulated on transcriptional level by both temperature and Ca^{2+}. Thus, at 37°C in a Ca^{2+} free medium these genes are heavily transcribed (Fig.3) (11,12). A number of insertion-mutants of the Ca^{2+} region of pIB1 have been isolated. These mutants fall into two different classes:One class is Ca^{2+}-independent (CI), whereas the other contains mutants that are temperature sensitive (TS) for growth at 37°C irrespective of the Ca^{2+} concentration of the medium. When these mutants were analyzed with respect to transcription of the *yopH* and *yopE* genes, the transcriptional activity was changed compared to their isogenic wild type strain YPIII(pIB103). The CI mutants were unable to transcribe the *yopE* and *yopH* genes during any of the conditions used, in good agreement with results obtained from Yop expression studies. The TS mutants showed a high level of transcription at 37°C either in the presence or absence of Ca^{2+} while no transcription was observed at 26°C (11,12). From these results we conclude that the expression of the Yops is regulated on the level of transcription by Ca^{2+} and temperature involving trans-acting elements encoded by the Ca^{2+} region.- The *yopE* promoter was cloned into the β-lactamase expression vector pJAC1 resulting in pYP51. Thus, the *yopE* promoter was fused to the *ampC* gene. pJAC1 is based on the pBR322 replicon and it has a high copy-number.Therefore, another plasmid construct was also made, pYP52, based on the pACYC184 replicon. This plasmid is identical to pYP51 with respect to ability to express β-lactamase but it has a lower copy-number

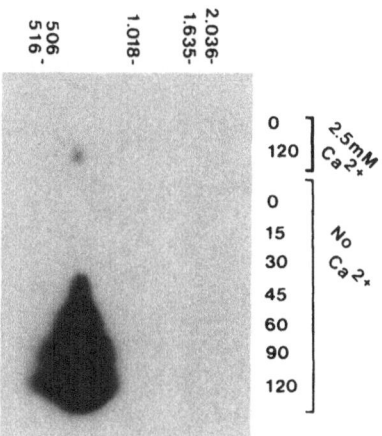

Fig. 3. Transcription of the *yopE* gene of strain YPIII (pIB1). RNA was extracted from cells sampled at indicated times (minutes) after a temperature shift from 26°C to 37°C.

compared topYP51 when measured in *Y. pseudotuberculosis*. The ratio in copy number was found to be around 10:1 for pYP51 versus pYP52 (11). These two plasmids were put into YPIII and YPIII(pIB1) respectively and the ability of these strains to express a-lactamase was analyzed (Fig.4). It was found that YPIII(pYP51) and YPIII(pYP52) expressed the enzyme at low levels irrespective of the incubation conditions in contrast to YPIII(pIB1,pYP52) that expressed a-lactamase at a high level at 37°C showing an elevated level of expression when Ca^{2+} was depleted (Fig.4). Strain YPIII(2pIB1,pYP51) expressed high levels of β-lactamase at 37°C irrespective if Ca^{2+} was present in the growth medium or not, while no expression was seen when the strain was incubated at 26°C (Fig.4). Two different conclusions can be drawn from these results: i) There is a pIB1 encoded trans-activator, that is able to stimulate the transcription of the *yopE* gene and this trans-activator is solely regulated by temperature; ii) There is, in addition, a plasmid-encoded repressor that regulates transcription of the *yopE* gene and the concentration of repressor is regulated by Ca^{2+}. To further dissect the regulation, the same type of experiments as described above was carried out but the wild-type strain was exchanged with different mutants of the Ca^{2+} region (Fig.5). One mutant YPIII(pIB61) *lcrI* was unable during any conditions to induce transcription of the *ampC* gene showing an activator-minus phenotype. This mutant did not show the same map-position as *lcrF* which recently was identified to be essential for trans--activation of the *yopH* gene of *Y. enterocolitica* (13). *lcrF* shows homology with AraC and it is likely to be an activator interacting with thepromoter regions of the *yop* genes (13). We have also identified thecorresponding gene *lcrF*, of strain YPIII(pIB1). This gene showed same localization in both *Y. enterocolitica* and *Y. pseudotuberculosis*. Thus, at least two different genes *lcIr* and *lcrF* respectively, are involved in the temperature activated transcriptional control of the Yops. When the negative control-loop was analyzed we observed

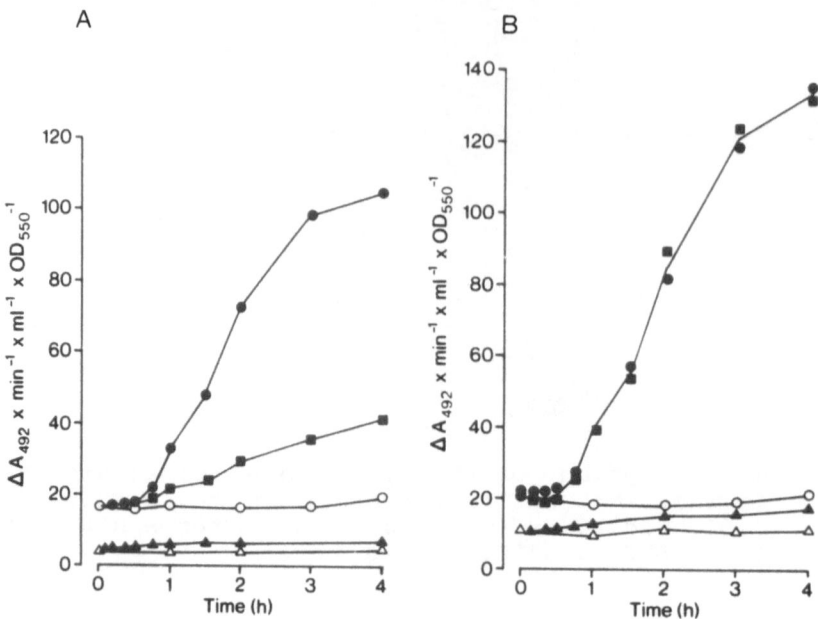

Figure 4. β-lactamase activity of two different *yopE* promoter hybrid plasmids, pYP51 and pYP52 of strains YPIII and YPIII(pIB1). Cell were grown at 26°C and subjected to a temperature shift to 37°C at time 0. Parallel cultures were kept at 26°C. A. Plasmid pYP52 promoter B. Promoter plasmid pYP51. △, YPIII 26°C. ▲, YPIII 37°C. ○, YPIII(pIB1) 26°C, ● YPIII-(pIB1) 37°C. ■, YPIII(pIB1) 37°C + 2.5 mM Ca^{2+}.

two different classes of mutants. One class was temperature sensitive for growth at 37°C irrespective of the Ca^{2+} concentration of the medium (11). This class of mutants showed the phenotype that would be expected from mutants affected in the actual repressor. Mutants in three different gene-loci, *lcrK*, *lcrE* and *lcrGVH-yopC-yopD* showed this behaviour, *i.e.* insertion mutants of these regions were TS and these mutants showed a high transcriptional activity even in the presence of Ca^{2+}. The second class of mutants was CI and transcription of the β-lactamase gene could only be obtained in constructs containing the high-copy number vector pYP51 (11). Such strains were in addition TS and Ca^{2+} had no effect on transcription. This class of mutants YPIII(pIB71), YPIII (pIB72) and YPIII(pYV019/2) was thus, affected in the ability to modulate the concentration of the repressor in response to the environmental Ca^{2+} signal, *i.e.* they showed a constitutive high concentration of repressor. We conclude from these results, that the negative-control-loop involves at least 5 genes, of which some are controlling the actual concentration of repressor in response to the environmental stimuli Ca^{2+}.

EXPORT OF THE YOPS

Two of the TS mutants affected in the *lcrK* gene, did not secrete the Yops to the culture medium although they showed a high transcriptional activity of the *yop* genes. This

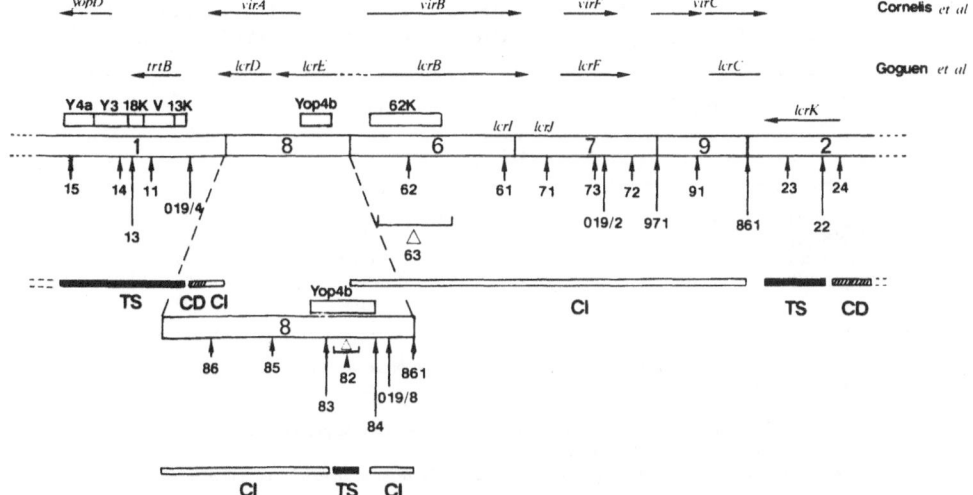

Fig. 5.   Insertion mutants of the Ca²⁺ region. The linear
           BamH1 map of the Ca²⁺ region of pIB1 is shown. The
           different insertion mutants are indicated by ar-
           rows. The numbers refer to the specific mutant.
           Above the map, data obtained from Goguen et.al.
           (16,17) and Cornelis et.al (13,18) on the corre-
           sponding regions of *Y. pestis* and *Y. enterocolitica*
           plasmids are shown. Horizontal arrows denote direc-
           tion of transcription. Abbreviations: CD, calcium
           dependent; CI, calcium independent, TS, temperature
           sensitive for growth at 37°C.

suggested to us that the Yops are secreted by a specific
plasmid-encoded mechanism. To test this hypothesis we first
constructed a *yopE* deletion mutant YPIII(pIB522) and introduced into this strain pTB1, which is able to express the
Yop5 protein from the *tac*-promoter after induction by IPTG.
It was observed that the Yop5 protein could only be recovered
from the culture supernatant after that strain YPIII(pIB522,
pTB1) had been incubated at 37°C in a Ca²⁺ free medium containing IPTG. If the strain was incubated at 26°C or if Ca²⁺
was added when incubated at 37°C no Yop5 could be detected in
the culture supernatant (Fig.6). Moreover, Yop5 was not
detected in strain YPIII(pTBI) irrespective of the incubation
conditions. These results indicated that the pIB1 encodes a
specific transport system for the secretion of the Yops, and
that this system is coregulated with the expression of the
Yops. However, are also unable to export the Yops. Thus, as
the regulation of expression the specific secretory system
seams to be very complex.Function of Yop5 and Yop2b in virulence Yop2b: The structural gene of Yop2b, *yopH* was recently
cloned and found to be located within 2.8 kb *PstI* fragment
(12). This fragment was sequenced and the only open-reading
frame that could be identified constituted the *yopH* gene. The
PstI fragment was deleted and replaced by a DNA-fragment
containing a kanamycin-resistant gene, generating pIB29 (12).
This mutant YPIII (pIB29) was found to be avirulent after
intraperitoneal and intravenous injection, having the same LD
value as the corresponding cured strain YPIII (Table 2).

Table 2. Virulence of various strains of *Y. pseudo-tuberculosis* after infection of mice via different routes.

Strain	Relevant genotypes	LD$_{50}$		
		I.V.[a] 14 d[c]	28d[c]	I.P.[b]
YPIII (pIB103)	wt	5×10^3	5×10^3	2×10^4
YPIII (pIB522)	*yopE*	2×10^6	1×10^5	9×10^6
YPIII (pIB540)	*yerA*	3×10^4	3×10^3	9×10^6
YPIII (pIB291)	*yopH*	$>10^6$	$>10^6$	1×10^7
YPIII	plasmid⁻	$>10^6$	$>10^6$	5×10^6

Mice were challenged with different doses of bacteria indicated in Table, either by;
a) intravenous injection (I.V.) or
b) intraperitoneal injection (I.P)
c) To illustrate the reduced infection rate between the wild-type strain and the mutants the LD$_{50}$ value was calculated 14 days and 28 days after infection.
The mice were recorded daily and the LD$_{50}$ value was established by the method of Reed and Muench (1938)

However when pYOP21, which is a hybrid plasmid based on pACYC184 carrying the 2.8 kb PstI fragment encoding Yop2b, was put into YPIII(pIB29) the resultant strain was found to be virulent.Thus, the Yop2b protein is an essential virulence determinant.Virulent *Y. pseudotuberculosis* has the ability to inhibit phagocytosis by mouse peritoneal macrophages in contrast to avirulent strains which lack this ability (14). Thus, the phagocytosis inhibition is linked to the virulence plasmid. The *yopH* deletion mutant YPIII(pIB29) showed a decreased ability to inhibit phagocytosis, while the trans--complemented strain YPIII(pIB29, pYOP21) had regained the ability to inhibit phagocytosis to the same extent as the wild type strain (Table 3). These results clearly demonstrate that Yop2b is involved in the molecular process that leads to inhibition of phagocytosis and that this ability is a virulence attribute. However, the level of inhibition of the *yopH* mutant was about 20% irrespective of the incubation conditions, while the corresponding value was 5% for the isogenic plasmid-cured strain YPIII, indicating that the intermediate level of inhibition is also dependent on the presence of the plasmid (Table 3). Thus, other plasmid-encoded products must also be involved in the ability of the pathogen to inhibit phagocytosis. Yop5: Mutants defective in the expression of Yop5 has also been constructed. Such mutants are also avirulent (11). During these studies a novel gene *yerA* was identified, which was found to map within a region of 620 bp upstream of the *yopE* gene. A *yerA* mutant, YPIII(pIB540), expresses lower amounts of Yop5 compared to its wild-type strain. Wild-type expression of Yop5 could be restored by trans-complementation, indicating that the 14 kD protein encoded by the *yerA* gene is a positive regulator of Yop5 expression. Moreover, a *yerA* mutant was also found to be avirulent after oral infection indicating that lowered expression of Yop5 is sufficient to reduce the virulence of the pathogen. We have earlier shown that the virulence-plasmid pIB1 encodes determinants essential for the ability of the pathogen to mediate a cytotoxic effect on confluent layer of HeLa cells (15). The*yopE* mutant YPIII(pIB522) was unable to

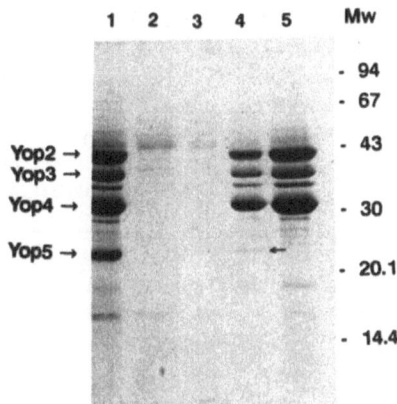

Fig. 6.   Secretion of Yop5 in strains having the *yopE* gene
under the control of the tac-promoter. The different strains were grown in a medium containing 2.5
mM Ca²⁺ or lacking Ca²⁺ at 26°C. IPTG, 1mM was
added to some of the cultures and thereafter the
cultures were shifted to 37°C. After 2 hours of
incubation the bacteria were harvested and the
resulting culture supernatant fluids obtained from
the respective strain was precipitated and subjected to SDS-PAGE. Lanes: 1: YPIII(pIB102) -Ca²⁺, 2:
YPIII(pTB1) -Ca²⁺ +IPTG, 3: YPIII(pIB522,pTB1)
+Ca²⁺ +IPTG, 4: YPIII(pIB522,pTB1) -Ca²⁺ +IPTG, 5:
YPIII(pIB522,pTB1) -Ca²⁺

mediate cytotoxicity. This defect could be trans-complemented
back to wild-type level by pAF55 which carries 1.5 kb. *BamHI*
/*ClaI* fragment encoding Yop5. A 0.9 kb *XhoI* fragment containing the whole coding sequence of the *yopE* gene without its
promoter, was spliced into the IPTG - inducible expression
vector pMMB66 to generate pTB1. When pTB1 was put into YPIII-
(pIB522) the resultant strain was able to mediate cytotoxicity only after the addition of IPTG. Together, these results
show that the Yop5 is required by the pathogen to induce a
cytotoxic effect on a confluent layer of growing HeLa cells.
As mentioned above a *yopH* mutant was still more resistant to
phagocytosis compared to the isogenic cured strain YPIII
(Table 3). Since the macrophages in these experiments also
were affected cytotoxically, it was suspected that the low
residual activity of the *yopH* mutant to inhibit phagocytosis
was linked directly to the cytotoxic effect. Thus, we constructed a *yopH/yopE* double mutant, YPIII(pIB251). This
construct was unable to inhibit phagocytosis by mouse peritoneal macrophages showing the same low level as YPIII (Table
3). In addition the double mutant was not cytotoxic for
macrophages in contrast to the respective single mutant
(Table 3). These results demonstrate that the Yop2b and Yop5
proteins actin concert in order to defeat the primary defence
mechanisms of the host. Therefore one main function of some
of the Yops in virulence seems to be to inhibit phagocytosis.
To further explore the role of the Yop5 protein in virulence

Table 3. Inhibition of phagocytosis by different strains of *Y. pseudotuberculosis*.

Strain	Relevant genotype	% extracellular bacteria of total MØ associated bacteria	Cytotoxic for MØ
YPIII	p^-	6 ± 2 (5)	-
YPIII(pIB29)	$yopH, yopE^+$	23 ± 3 (5)	+
YPIII(pIB29,pYOP21)	$yopH^+, yopE^+$	63 ± 4 (5)	+
YPIII(pIB522)	$yopH^+, yopE$	54 ± 7 (3)	+
YPIII(pIB251)	$yopH, yopE$	7 ± 3 (3)	-

we performed new virulence tests including intraperitoneal (I.P.) and intravenous (I.V.) infections by using two mutants affected in the expression of Yop5. After I.P. infection, mutants YPIII(pIB522) *yopE* and YPIII(pIB540) *yerA* were as avirulent as the plasmid-cured strain YPIII (Table 2). However following I.V. infection both mutants were more virulent compared to YPIII (Table 2). YPIII(pIB540) showed a LD value which approached that of the corresponding wild-type strain YPIII(pIB103) (Table 2); the LD value of YPIII(pIB522) was intermediate to the values of YPIII and YPIII(pIB103) (Table 2). Thus, when these mutants are allowed to pass, by artificial means, a barrier of the host defence they are virulent. These results demonstrated that a important function of the Yop5 protein in the virulence process is to obstruct the antibacterial activities of the professional phagocytes of the lymph nodes. Reduced or abolished expression of Yop5 would then lead to attenuated strains having a lowered ability to resist the host defense and consequently such bacteria would be eliminated in the regional lymphnodes.

CONCLUSIONS

- All three virulent species of *Yersinia* contain homologous virulence plasmids that encode a number of temperature-inducible secreted/outer-membrane proteins (Yops).

- Expression of Yops is regulated at the level of transcription involving negative as well as positive control elements encoded by the Ca^{2+} region.

- The activator is controlled by temperature and the repressor is controlled by the concentration of calcium.

- Yop2b and Yop5 are essential virulence determinants.

- Yop2b and Yop5 obstruct the primary host-defense at level of inhibition of phagocytosis by professional phagocytes.

ACKNOWLEDGEMENTS

This work was supported by the Medical Research Counsil of Sweden.

REFERENCES

1. Portnoy, D.A., H. Wolf-Watz, I. Bolin, A.B. Beeder, and S. Falkow. 1984. Infect.Immun. 43:108-114.
2. Kupferberg, L. and K. Higuchi. 1958. J.Bacteriol. 76:120-121
3. Brubaker, R.R. 1967. J. Infect. Dis. 117:403-417.
4. Gemski, P.K., J. Lazere, and T. Casey. 1980. Infect. Immun. 27:682-685.
5. Forsberg,]., I. Bolin, L. Norlander, and H. Wolf-Watz. 1987. Microbial Pathogen. 2:123-137.
6.   Bolin, I.,]. Forsberg, L. Norlander, M. Skurnik, and H. Wolf-Watz. 1988. Infect.Immun. 56:343-348.
7. Wolf-Watz. H., I. Bolin,]. Forsberg, and L. Norlander. 1986. In D. Lark ed Academic PressLondon 329-334.
8. Sodeinde, O.A., A.K. Sample, R.R. Brubaker, and J.D. Goguen. 1988. Infect. Immun. 56:2749-2752.
9. Bolin, I., D. Portnoy, and H. Wolf-Watz 1985. Infect. Immun. 48:234-240.
10. Perry, R.D., P.A. Harmon, W.S. Bowmer, and S.C. Straley. 1986. Infect. Immun. 54:428-434.
11. Forsberg,]. and h. Wolf-Watz. 1988. Mol. Microbiol. 2: 121-133.
12. Bolin, I. and H. Wolf-Watz. 1988. Mol. Microbiol. 2: 237-245.
13. Cornelis, G., C. Sluiters, C. Lambert deRouvroit and T. Michels. 1989. J.Bacteriol. 171:254-262.
14. Rosqvist, R. and H. Wolf-Watz. 1988. Infect. Immun. 56: 2139-2143.
15. Rosqvist, R. and H. Wolf-Watz. 1986. Microbiol Pathogen. 1:229-240.
16. Goguen, J.D., J. Yother and S. Straley. 1984. J. Bacteriol. 160:842-848.
17. Yother, J., T. Chamness and J.D. Goguen. 1986. J. Bacteriol. 165:443-447.
18. Cornelis, G., M.P. Sory, Y. Laroche and I. Derclaye. 1986. Microb. Pathogen 1:349-359.

OUTER MEMBRANE PROTEINS: OLD AND NEW

Peter Owen, Patrick Caffrey[1], Lars-Goran Josefsson[2] and Mary Meehan

Department of Microbiology, Moyne Institute, Trinity College, Dublin, Ireland

SOMETHING OLD

The outer membrane of *Escherichia coli* has been studied extensively for almost two decades and much is now known about the composition, structure, function, immunology, biosynthesis and regulation of the proteins which contribute to this outer protective layer (for reviews see refs 1-10). It is known to contain a number of major protein species, several of which e.g. the OmpF, OmpC, PhoE, LamB, Tsx proteins and Protein K are porins involved in the passive accumulation of small hydrophilic solutes. Others, like the iron-regulated outer membrane proteins (the *fepA*, *fhuA*, *fhuE* and *fecA* gene products) and the vitamin B_{12} receptor (the BtuB protein) are involved in the TonB-dependent uptake of more specific solutes (11-13). Other major proteins appear to play a structural role for the cell envelope. Falling into this category are proteins such as the Braun lipoprotein (the Lpp protein), peptidoglycan-associated lipoprotein (PAL) and the *ompA* gene product. Several other minor or less well characterized proteins, such as phospholipase A, the OmpT protein (a trypsin-like protease), signal peptidase, the *tolC* gene product (possibly involved in protein processing), Protein III, and an 83-kilodalton (kDa) iron regulated protein have also been identified (2,4,11)

There is emerging evidence that certain outer membrane proteins may be considered as virulence determinants (14). However, it should be stated that the above-mentioned chromosomally-encoded outer membrane proteins do not appear to play a major role in disease processes caused by *E. coli*. Rather, certain additional plasmid-encoded outer membrane proteins have been implicated in this respect. For other pathogens, however, this may not be the case. For example, in *Neisseria gonorrhoea*, there is evidence that chromosomally-encoded proteins, analogous to some of the major proteins of *E. coli*, do play an important role in virulence.

---

[1] Present address: Department of Biochemistry, University of Cambridge, Cambridge CB2 1QW, United Kingdom.
[2] Present address: Department of Cell Research, Swedish Agricultural University, S-751 24, Uppsala, Sweden

Microbial Surface Components and Toxins in Relation to Pathogenesis
Edited by E.Z. Ron and S. Rottem, Plenum Press, New York, 1991

127

Outer membrane proteins can be considered to facilitate infection of and survival within the host in at least four different ways (14). First, they may be involved in the acquisition of essential iron. In this respect, invasive strains of *E. coli* make extensive use of the aerobactin-mediated iron uptake system with its highly specific 74-kDa receptor (IutA protein) in the outer membrane. Indeed, there is now convincing genetic evidence that IutA represents a major virulence determinant for these organisms (15-20). Another example is the 86-kDa outer membrane protein (termed OM2) which acts as the iron-siderophore receptor for an efficient plasmid-encoded iron scavenging system in the fish pathogen *Vibrio anguillarum* (21-23). In addition, some human pathogens, notably *Neisseria meningitidis, N. gonorrhoea, Bordetella pertussis* and *Haemophilus influenzae* may possess outer membrane proteins which are capable of directly mobilizing iron from iron-binding proteins of the host e.g. transferrin, lactoferrin, and hemoglobin (14, 24-26).

It is generally recognized that fimbriae/pili play a key role in the ability of an organism to bind to epithelial cells (27). However, this may not be the whole story, as some outer membrane proteins have also been heavily implicated in this process. For example, the heat-modifiable and antigenically diverse Opa proteins (formerly Protein IIs or P.IIs; M_rs 24,000-32,000) of *N. gonorrhoea*, the plasmid-encoded YOP1 protein of *Yersinia enterocolitica* (M_r=47,000), and enteroadherence factor (EAF; M_r 94,000) encoded by the 60-MDa virulence plasmid of enteropathogenic *E. coli* (EPEC) have all been correlated with the ability of the corresponding pathogen to adhere to epithelial tissue (28-39).

Outer membrane proteins are also becoming increasingly implicated in the ability of some pathogenic microorganisms e.g. the gonococcus and species of *Escherichia, Shigella, Salmonella,* and *Yersinia* causing invasive diarrhoea, to enter epithelial cells. Notable examples in this respect include invasion (the 105 kDa *inv* gene product of *Yersinia pseudotuberculosis* (40-42]), the plasmid-encoded IpaA-D proteins (approximate M_rs 78,000, 57,000, 43,000 and 39,000, respectively) of *Shigella flexneri* and enteroinvasive *E. coli* (43-46) and the major gonococcal porin, Por (formerly Protein I or P.I; M_r between 32,000 and 39,000; refs. 32, 47).

A fourth mechanism by which outer membrane proteins may contribute to virulence is in evasion of the immune system. Well documented strategies in this respect include antigenic and phase variation of the type exhibited by neisserial Opa proteins (48- 50), production of blocking antibodies by the 31-kDa, 2- mercaptoethanol-modifiable gonococcal Rmp protein (formerly Protein III or P.III; refs. 32,51, 52), resistance to phagocytosis e.g. the Opa proteins of *N. gonorrhoea* (30, 53), and increased resistance to serum killing. The plasmid encoded Tra T lipoprotein (M_r 25,000) of *E. coli* and the gonococcal Por protein (P.IA) may be examples which fall into this last category (14, 47, 54, 55).

It could also be argued that proteins located in the outer membrane and facilitating the anchorage or secretion of virulence determinants have themselves crucial roles to play in pathogenesis. In this respect, consideration should be given to proteins such as the HlyB/D proteins (M_rs 66,000 and 53,000) which specifically facilitate the passage of the 107-kDa haemolysin (HlyA) protein through the *E. coli* envelope (56,57) and to minor pilin subunits such as the *faeD* and *fanD* gene products (M_rs 82,000 and 85,500) which appear to anchor the parent K88ab and K99 fimbriae, respectively, in the outer membrane (58).

Although they are rarely considered as true outer membrane components, proteinaceous surface structures such as S-layers, fimbriae and flagella all have an association and indeed often cofractionate with the outer membrane. Whereas the roles of fimbriae and flagella in adhesion and motility/-chemotaxis, respectively, have been well documented (27, 59), the functions of S-layers are less clear (60-62). S-layers have been detected in a wide range of bacteria, notable exceptions being members of the *Enterobacteriaceae*. They can usually be observed by electron microscopy as regular lattice structures covering the entire cell surface. Generally S-layers are composed of a single (glyco)protein species (M_rs between 40,000 and 200,000) but occasionally stoichiometric amounts of more than one protein species have been observed. Proposed roles for S-layers are diverse and often speculative and include roles in cell shape determination, protection, molecular sieving, ion trapping, cell adhesion and surface recognition (60-62). In one case, viz for the fish pathogen *Aeromonas salmonicida,* there is strong evidence for a role in virulence. Here, the structure (termed A-layer) appears to facilitate the spread of infection by imparting resistance to the bacteriocidal activity of serum complement, by enhancing the ability of the organism to associate with macrophage and to resist effects of proteases, and possibly in the sequestration of iron (63-66).

SOMETHING NEW

Most of the outer membrane proteins listed above were identified initially following analysis of membrane fractions by sodium dodecyl sulphate polyacrylamide gel electrophoresis (SDS-PAGE), by electron microscopy, after genetic manipulation or by a combination of these approaches. A number of years ago, following the introduction of the high resolution, two dimensional non-denaturing technique of crossed immunoelectrophoresis (CIE), workers in this laboratory began a systematic immunochemical analysis of the envelopes of *E. coli* ML308-225 (reviewed in refs. 67-69). During the course of these studies, a novel bipartite protein was identified. The remainder of this article will be devoted to a review of the properties of this somewhat unique outer membrane antigen.

Preliminary studies by crossed immunoelectrophoresis

In the late 1970's, CIE studies of membrane vesicles and envelopes of aerobically-grown *E. coli* succeeded in establishing a reference profile containing over 50 well-resolved immunoprecipitates. The immunoprecipitates were numbered for convenience in order of decreasing electrophoretic mobility i.e. lower numbers corresponding to more acidic antigens (70). The combined information obtained from the subsequent application of a variety of related immunochemical procedures (including use of purified antigens or monospecific antisera or mutants lacking specific gene products, charge-shift immunoelectrophoresis, zymograms, analysis for specific [radiolabeled] cofactors, polypeptide analysis, and immuno-absorption) resulted in the identification in functional terms or partial characterization of over half of the component antigens (67-77). Thus, major envelope-associated antigens such as ATPase, succinate dehydrogenase, NADH dehydrogenase, the OmpF/C and OmpA proteins, the Braun lipoprotein and lipopolysaccharide (LPS) were readily identified. Importantly and unexpectedly, CIE analysis of purified radiolabeled outer membranes (76) demonstrated the existence of a prominent immunoprecipitate (no 43), analysis of which revealed the presence in equal stoichiometry of two (radiolabeled) polypeptides of approximate M_rs 60,000 (termed α) and 53,000 (termed β).

The detection of such a bipartite (heterooligomeric) antigen was interesting since it implied a complexity of organization not normally associated with proteins of the outer membrane. It will be recalled that most outer membrane proteins described to date appear to exist as monomers or as homooligomers (e.g. homotrimers). There is some evidence from cross-linking experiments for interactions between the OmpA protein and the Braun lipoprotein (2) but a complex of defined stoichiometry has not been isolated. There is also some evidence that the closely- related OmpF and OmpC proteins can form mixed (hetero)trimers in situ (2,78). Perhaps the strongest evidence for a heteropolymeric complex of two outer membrane proteins occurs in *N. gonorrhoea*, where several groups have reported an association between the Por and the Rmp proteins (79-81). For this and other reasons, antigen 43 became the focus of more intensive study.

Polypeptide composition

High-titre antiserum specific for antigen 43, and raised by immunization with immunoprecipitates excised from CIE immunoplates, was used to facilitate the analysis by SDS-PAGE of the immune complex obtained following coincubation of serum with Triton X-100 extracts of ^{14}C-amino-acid labeled envelopes of *E. coli* (82). The results obtained confirmed and extended earlier observations obtained by precipitate excision (76). Again two polypeptides were observed in equal stoichiometry. The α subunit (apparent M_r 60,000) had an electrophoretic mobility which was unaffected by the temperature (50-100°C) used to solubilize the complex. The other subunit was heat modifiable in a manner similar to that of several other outer membrane proteins (2), and migrated with apparent M_rs of 37,000 (β')if heated to temperatures of 70°C or below and 53,000 (β) if heated at higher temperatures. The α and β polypeptides did not appear to be linked by disulfide bonds. Identical behaviour was also observed for the corresponding polypeptides which were often prominent features of the SDS-PAGE profile of isolated envelopes or outer membrane preparations.

The α and β polypeptides were clearly different proteins since they gave radically different peptide profiles following treatment with V8 protease and generated non-cross reactive, subunit specific antisera following immunization. Furthermore, immunoprecipitation experiments performed on envelopes prepared from *E. coli* ML308-225 grown in the presence of [^{14}C]succinate as sole carbon source failed to detect any trace of the ladder profiles characteristic of the heterogeneously-sized smooth LPS of this organism. Thus antigen 43 seemed to be composed of two chemically and immunologically distinct protein subunits in 1:1 ratio i.e. [α β]$_x$ (82).

Location of antigen 43

In view of the unique composition of antigen 43, it became important to establish conclusively its location in the cell envelope. Analysis of inner and outer membranes separated by isopycnic centrifugation or following treatment with Triton-100-Mg^{2+} clearly revealed that both polypeptides of antigen 43 partitioned in a manner identical to that observed for known marker proteins of the outer membrane. Nor could either polypeptide be detected by Western blotting in isolated inner membranes, soluble (cytoplasm/periplasm) fractions of the cell, or in concentrated cell-free supernatants (82).

Progressive immunoadsorption experiments conducted several years ago had established that antigen 43, together with a limited number of other components, were accessible to antibody on the surface of *E. coli* ML308-225 (72). These results were then extended by Western blot analysis which established that intact cells adsorb from antimembrane serum antibodies to the α subunit but not to the β subunit. More recent immunofluorescence studies and immunogold labelling experiments performed on thin sections of *E. coli* also clearly locate antigen 43 to the outer membrane (Meehan, Beesley and Owen, unpublished data). From all these data, there seems little doubt that antigen 43 is a component of the outer membrane and that the α subunit is surface expressed.

Do the α and β polypeptides form a complex in situ?

Several lines of evidence lent support to the hypothesis that the α and β polypeptides represent subunits of a novel (bipartite) protein antigen. First, the two polypeptides were clearly present in equal stoichiometry. Second, the α and β polypeptides copartitioned and maintained this stoichiometry following a range of nondenaturing treatments. Third, whereas the two proteins in question were chemically and immunologically distinct, test tube immunoprecipitation reactions conducted under non-denaturing conditions with Triton X-100-solubilized envelopes and subunit- specific antisera always resulted in coprecipitation of similar amounts of the α and β polypeptides. This was in marked contrast to results obtained using denatured membranes. In these cases, the two antisera only reacted with their homologous subunit (82). In more recent experiments (83), reconstitution of the αβ complex from separated subunits has been observed. This was achieved by mixing purified α subunit with Triton X-100-EDTA solubilized outer membranes which had been specifically depleted of the α subunit by heat treatment. Thus, whereas anti-β antibody did not precipitate purified α subunit and anti-α serum failed to precipitate the β subunit from α-depleted outer membranes, each antisera induced coprecipitation of both α and β subunits from mixtures obtained by preincubating free α subunit with α-stripped outer membranes (83).

Properties of the α and β subunits

A key observation in the analysis of antigen 43 was the finding that the α but not the β subunit was selectively and almost quantitatively released from *E. coli* outer membranes by brief heating to 60°C (83). Provided clean outer membranes or well washed Triton X-100-Mg^{2+}-extracted envelopes were used, then the α subunit constituted over 80% of the total protein released. A number of procedures which failed to release the α subunit included treatment of outer membranes with 2 M urea, 4 M guanidine hydrochloride, 2 M NaCl or 5% (vol/vol) 2-mercaptoethanol. Purification to apparent homogeneity could be achieved by gel filtration on Sephacryl S200/-S300 superfine. The subunit eluted as a monomer of M_r 50,000 and gave a single band (apparent M_r 60,000) when analyzed by SDS-PAGE and staining by either Coomassie blue or silver nitrate. Immunological tests indicated the presence of a single antigenic species and analysis by the silver-staining technique the presence of less than 1% LPS (83).

Heat-stripped outer membranes provided a useful starting point for the purification of the β subunit since it could be shown that the latter retained its properties of heat-modifiability following release of the α polypeptide. Detergents were required to solubilize the β subunit, however, the latter behaving in this respect much like other outer membrane proteins. Purification was finally achieved by combination of gel filtration in presence of SDS, cold butanol precipitation, and preparative SDS-PAGE. When analyzed by SDS-PAGE, the final product ran as a single protein staining band which retained its properties of heat modifiability, and contained undetectable levels of LPS (Caffrey, Meehan and Owen, unpublished data).

The α subunit had a pI of 4.6, a polarity of 49.2%. and an amino acid composition which was notable for the abundance of glycine (102 residues/mole). The β subunit had a polarity of 47.9%. Neither subunit contained cysteine residues or appeared to be glycosylated (83; Caffrey, Meehan and Owen, unpublished data). CIE performed in conjunction with zymogram stains or analysis for specific (radiolabeled) cofactors viz $^{59}Fe^{3+}$, [^{14}C]riboflavin, [^{3}H]palmitate or [^{3}H]glycerol revealed that antigen 43 neither expressed any of a wide variety of enzyme activities (listed in ref. 67) nor possessed iron or flavin cofactors. Neither polypeptide was soluble in aqueous organic solvents nor labeled with [^{3}H]palmitate or [^{3}H]glycerol in a manner typical of envelope-associated lipoproteins (77 and refs. therein). Nor did either subunit appear to be peptidoglycan associated. From an analysis of the amino acid compositions, N-terminal sequences, behavior on SDS-PAGE gels, defined mutants, and reactions with specific antisera it was possible to rule out a relationship between either polypeptide and other documented *E. coli* protein antigens (e.g. flagellin and soluble common protein antigen, the BtuB protein, the TolC protein and OmpA protein), which possessed similar M_r or properties and were capable of an association with outer membrane fractions (82).

No significant homology was found between the N-terminal sequence of the β subunit and other published sequences (Caffrey, Meehan and Owen, unpublished data). However, the N-terminus of the α subunit showed some notable similarities with sequences in the N-terminal regions of certain fimbriae of *E. coli* (83). It had been noted previously (84) that a wide range of fimbrial subunits contained two invariant glycine residues (corresponding to positions 8 and 14 in the *E. coli* type 1 fimbrial subunit). Interestingly, the primary structure of antigen 43α subunit also conformed to this pattern (Fig. 1). More strikingly, the 43α sequence contained a stretch of six residues (TVNGGT) which was also present in the N-termini of the major subunits of type 1 fimbriae and the related F1C (KS71C) and S fimbriae. Indeed, alignment of the sequences of 43α (residues 10-20 inclusive) and the *Klebsiella* type 1 fimbrial subunit reveals that at 8 positions in this 11-residue region the amino acids are identical. At two of the remaining three positions the aligned residues are isofunctional (83; see Fig. 1). The precise significance of these observations is unclear at the present time. However, the polypeptides in question all belong to a limited number of proteins which are exported through the outer membrane. It has been suggested that conserved sequences in the N-terminal domains of enterobacterial fimbrial subunits may play a role in export and assembly of fimbrial organelles (85).

Electron microscopy

Selective detachment from cell envelopes is typical of appendages such a fimbriae, flagella and S-layers (86, 87). However, electron microscopic studies of negatively-stained preparations of cells, purified outer membranes, free α subunit or immunoprecipitated antigen 43 did not reveal a morphologically recognizable or regular repeating structure. It should be noted, perhaps, that some S-layers and fimbrial structures have proved exceptionally difficult to visualize in the electron microscope (60, 61, 86, 88). Therefore, it may be that the use of more sophisticated methodology is required in order to observe the quaternary structure of the complex.

Antigen 43 undergoes phase variation

A notable and often frustrating feature of early studies was the fluctuating yields obtained for antigen 43 following growth of *E. coli* under rigidly standardized conditions (82). In some cultures the levels would be high, with copy numbers approaching that of other major outer membrane proteins such as the OmpF/C and OmpA proteins. In others, only trace amounts of the antigen might be detected. This was in sharp contrast to the behaviour of most other outer membrane proteins. Furthermore, it became clear that the expression of neither subunit appeared to be dramatically dependent on growth conditions per se. This again is in marked contrast to the behaviour of other outer membrane proteins such as the OmpF, OmpC, LamB proteins and iron chelate receptors, for example, which respond to changes in osmolarity, and maltose and iron concentrations in predictable and often dramatic fashions (2,5).

```
                      *  *  *  *  *  *              *
43α        A D N V K H P W E T V N G G T L A G H G N Q - G F G

Type 1ᴱ             A A T T V N G G T V H F K G E V V N A A

FIC                 V T T V N G G T V H F K G E V V N A A

S                   V T T V N G G T V H F K G E V V D A A

Type 1ᴷ             A T T V N G G T V A F K G E V V D A A

Type 1ˢ        A D P T P V S V S G G T I H P E G K L V N A A
```

Fig. 1.   Comparison of the N-terminal seqence of antigen 43α
          subunit with the N-termini of various fimbrial
          subunits (data from refs. 84, 85, 95-98). Type 1ᴱ,
          type 1ᴷ and type 1ˢ fimbriae are from *E. coli*, *K.
          pneumoniae* and *Salmonella typhimurium*, respective-
          ly. Amino acids are given in the standard single
          letter code. Dashes indicate residues which were
          not identified, and asterisks those which are con-
          served. All sequences begin with residue 1 of the
          mature protein.

Phase variation was later found to be responsible for
this phenomenon (83). Antigen 43+ and antigen 43- variants
could readily be distinguished by filter immunoassays of
colonies on solid media, SDS-PAGE analysis of outer membranes
confirming the much decreased levels of both subunits from
the antigen 43- variants. Screening of progeny derived from
both positive and negative variants indicated that variation
was reversible, the rate of phase variation from antigen 43-
to antigen 43+ states approximating to 6 x 10⁻⁴ (82). More
recent immunofluorescence experiments have confirmed these
results and have indicated that most (but not all) cells
derived from antigen 43+ colonies express the antigen on
their suface, whereas only a minority of cells derived from
antigen 43- colonies do so (Caffrey, Meehan and Owen, unpublished datea). To date there is no evidence for different
antigenic forms of the complex. Thus, the phenomenon appears
to be similar to that characterized by type 1 fimbriae, i.e.
one of reversible phase variation and not antigenic variation
(89, 90).

Antigen 43 has been detected by Western blotting in a
wide variety of *E. coli* strains, both rough and smooth. In
some, notably ML308-225, C600 and BRE69, the two polypeptides
could be readily detected in the Coomassie blue stained
profile of outer membranes resolved by SDS-PAGE. In others,
the polypeptides could only be discerned by Western blotting
(82). So far we have no evidence that the antigen is expressed in organisms other than *E. coli*. However, the phenomenon of phase variation precludes definitive statements in
this regard until such time as a DNA probe becomes available.

Function

The balance of the available data strongly support a

working model in which antigen 43 is composed of two chemically and immunologically distinct protein subunits in 1:1 ratio. At present, we visualize the α subunit as extending through the O-antigen chains of LPS and being associated with the outer membrane through its interaction with the β subunit, an integral membrane protein.

As yet, we are unable to ascribe a function to antigen 43. The most obvious candidate in view of its surface expression, N-terminal sequence, and phase variability would be a novel type of adhesin. The M_r of the α subunit is, of course, far outside the normal range (14,000 to 29,000) for the major subunits of *E. coli* fimbriae (84, 86). However, it is possible that in the case of antigen 43 α subunit, an adhesive domain and a structural domain reside within one polypeptide chain. Levine et al. reported that a 94,000 dalton plasmid-encoded outer membrane protein may constitute part of an adherence factor of enteropathogenic *E. coli* strains (38, 39). It will be interesting to compare this protein with antigen 43 when more data become available.

It is also conceivable that antigen 43 is a type of S-layer. The α subunit is readily detached from the cell surface as are S-layers (60-62). Indeed, antigen 43 bears a passing resemblance to the *Caulobacter crescentus* S-layer which contains a 130-kDa subunit associated with a 71-kDa subunit itself embedded in the outer membrane (91). However, as has been mentioned earlier, crystalline surface arrays have never been detected among members of the *Enterobacteriaceae*.

Antigen 43 stimulates interest from an additional viewpoint. The α subunit is one of few proteins on the *E. coli* surface which is readily accessible to antibody. Furthermore, it can be easily purified in undenatured state. In these respects, it may prove to be an ideal carrier protein for the genetic insertion of foreign epitopes. Similar approaches using LamB, OmpA and P-fimbriae have been adopted with a view to developing new diagnostic tests or vaccines (92-94).

REFERENCES

1. Arbuthnott, J.P., Owen, P. and Russell, R.J. 1983. in Topley and Wilson's Principles of Bacteriology, Virology and Immunity, 7th Edition, Wilson, G. and Dick, H.M., Eds., Vol.1, pp. 337-373, Edward Arnold, London.
2. Lugtenbeg, B. and van Alphen, L. 1983. Biochim. Biophys. Acta 737, 51-115.
3. Benz, R. 1985. CRC Crit. Rev. Biochem. 19, 145-190.
4. Lugtenbeg, B. 1985. in Enterobacterial Surface Antigens: Methods for Molecular Characterization Korhonen, T.K., Dawes, E.A. and Makela, P.H., Eds., pp. 3-16, Elsevier Science Publichers B.V., Amsterdam.
5. Nikaido, H. and Vaara, M. 1985. Microbiol. Rev. 49, 1-32.
6. Oliver, D. 1985. Annu. Rev. Microbiol. 39, 615-648.
7. Owen, P. 1985. in Enterobacterial Surface Antigens: Methods for Molecular Characterization Korhonen, T.K., Dawes, E.A. and Makela, P.H., Eds., pp. 207-242, Elsevier Science Publishers B.V., Amsterdam.

8. Smyth, C.J. 1985. in Immunology of the Bacterial Cell Envelope Stewart-Tull, D.E.S. and Davis, M., Eds., pp. 177- 201, John Wiley and Sons Ltd., London.
9. Hancock, R.E.W. 1987. J. Bacteriol. 169, 929-933.
10. Inouye, M. 1987. Bacterial Outer Membranes as Model Systems, pp.1-450, John Wiley and Sons Ltd., New York.
11. Neilands, J.B. 1982. Annu. Rev. Microbiol. 36, 285-309.
12. Postle, K. and Good, R.F. 1983. Proc. Natl. Acad. Sci. U.S.A. 80, 5235-5239.
13. Braun, V. 1985. Trends Biochem. Sci. 10, 75-78.
14. Owen, P. 1988. in Immunochemical and Molecular Genetic Analysis of Bacterial Pathogens Owen, P. and Foster, T.J. Eds., pp. 27-43, Elsevier Science Publishers B.V., Amsterdam.
15. Montgomerie, J.Z., Bindereif, A., Neilands, J.B., Kalmanson, G.M. and Guze, L.B. 1984. Infect. Immun. 46, 835-838.
16. Carbonetti, N.H., Boonchai, S., Parry, S.H., Vaisanen--Rhen, V., Korhonen, T.K. and Williams, P.H. 1986. Infect. Immun. 51, 966-968.
17. Bagg, A. and Neilands, J.B. 1987. Microbiol. Rev. 51, 509- 518.
18. LaFont, J.-P., Dho, M., D'Hauteville, H.M., Bree, A., and Sansonetti, P.J. 1987. Infect. Immun. 55, 193-197.
19. Linggood, M.A., Roberts, M., Ford, S., Parry, S.H. and Williams, P.H. 1987. J. Gen. Microbiol. 133, 835-842.
20. Johnson, J.R., Moseley, S.L., Roberts, P.L. and Stamm, W.E. 1988. Infect. Immun. 56, 405-412.
21. Crosa, J.H. 1984. Annu. Rev. Microbiol. 38, 69-89.
22. Actis, L.A., Potter, S.A. and Crosa, J.H. 1985. J. Bacteriol. 161, 736-742.
23. Wolf, M.K. and Crosa, J.H. 1986. J. Gen. Microbiol. 132, 2949-2952.
24. Schryvers, A.B. 1988. Mol. Microbiol. 2, 467-472.
25. Schryvers, A.B. and Morris, L.J. 1988. Infect. Immun. 56, 1144-1149.
26. Schryvers, A.B. and Morris, L.J. 1988. Mol. Microbiol. 2, 281-288.
27. Smyth, C.J. 1988. in Immunochemical and Molecular Genetic Analysis of Bacterial Pathogens Owen, P., and Foster, T.J. Eds., pp. 13-25, Elsevier Science Publishers B.V., Amsterdam.
28. Zak, K., Diaz, J.-L., Jackson, D. and Heckels, J.E. 1984. J. Infect. Dis. 149, 166-174.
29. Schwalbe, R.S., Sparling, P.F. and Cannon, J.G. 1985. Infect. Immun. 49, 250-252.
30. Heckels, J.E. 1986. in Antigenic Variation in Infectious Diseases Birkbeck, T.H. and Penn, C.W., Eds., pp. 77-94, IRL Press, Oxford.
31. Bessen, D. and Gotschlich, E.C. 1987. Infect. Immun. 55, 141-147.
32. Blake, M.S. and Gotschlich, E.C. 1987. in Bacterial Outer Membranes as Model Systems Inouye, M., Ed., pp. 377-400, John Wiley and Sons Ltd., New York.
33. Balligand, G., Laroche, Y. and Cornelis, G. 1985. Infect. Immun. 48, 782-786.
34. Cornelis, G., Laroche, Y., Balligand, G., Sory, M.-P. and Wauters, G. 1987. Rev. Infect. Dis. 9, 64-87.
35. Heesemann, J. and Gruter, L. 1987. FEMS Microbiol. Lett. 40, 37-41.
36. Kapperud, G., Namork, E., Skurnik, M. and Nesbakken, T. 1987. Infect. Immun. 55, 2247-2254.

37. Forsbeg, A. and Wolf-Watz, H. 1988. Mol. Microbiol. 2, 121- 133.
38. Levine, M.M., Nataro, J.P., Karch, H., Baldini, M.M., Kaper, J.B., Black, R.E., Clements, M.L. and O'Brien, A.D. 1985. J. Infect. Dis. 152, 550-559.
39. Levine, M.M. 1987. J. Infect. Dis. 155, 377-389.
40. Isberg, R.R. and Falkow, S. 1985. Nature 317, 262-264.
41. Isberg, R.R., Voorhis, D.L. and Falkow, S. 1987. Cell 50, 769-778.
42. Small, P.L.C., Isberg, R.R. and Falkow, S. 1987. Infect. Immun. 56, 1674-1679.
43. Hale, T.L., Oaks, E.V. and Formal, S.B. 1985. Infect. Immun. 50, 620-629.
44. Oaks, E.V., Hale, T.L. and Formal, S.B. 1986. Infect. Immun. 53, 57-63.
45. Sansonetti, P.J., Ryter, A., Clerc, P., Maurelli, A.T. and Mounier, J. 1986. Infect. Immun. 51, 461-469.
46. Buysse, J.M., Stover, C.K., Oaks, E.V., Venkatesan, M. and Kopecko, D.J. 1987. J. Bacteriol. 169, 2561-2569.
47. Judd, R.C. 1989. Clin. Microbiol. Rev. 2, S41-S48.
48. Stern, A., Nickel, P., Meyer, T.F. and So, M. 1984. Cell, 37, 447-456.
49. Stern, A., Brown, M., Nickel, P. and Meyer, T.F. 1986. Cell, 47, 61-71.
50. Stern, A. and Meyer, T.F. 1987. Mol. Microbiol. 1, 5-12.
51. Blake, M.S., Wetzler, L.M., Gotschlich, E.C. and Rice, P.A. 1989. Clin. Microbiol. Rev. 2, S60-S63.
52. Rice, P.A. 1989. Clin. Microbiol. Rev. 2, S112-S117.
53. Virji, M. and Heckels, J.E. 1986. J. Gen. Microbiol. 132, 503-512.
54. Binns, M.M., Mayden, J. and Levine, R.P. 1982. Infect. Immun. 35, 654-659.
55. Aguero, M.E., Aron, L., DeLuca, A.G., Timmis, K.N. and Cabello, F.C. 1984. Infect. Immun. 46, 740-746.
56. Mackman, N., Nicaud, J.-M., Gray, L. and Holland I.B. 1986. Curr. Topics Microbiol. Immunol. 125, 159-181.
57. Gerlach, J.H., Endicott, J.A., Juranka, P.F., Henderson, G., Sarangi, F., Deuchars, K.L. and Ling, V. 1986. Nature 324, 485-489.
58. Mooi, F.R., Roosendaal, B., Oudega, B. and de Graaf, F.K. 1986. in Protein-Carbohydrate Interactions Lark, D.L., Ed., pp. 19-26, Academic Press Inc., London.
59. Smyth, C.J. 1988. in Immunochemical and Molecular Genetic Analysis of Bacterial Pathogens Owen, P. and Foster, T.J., Eds., pp. 3-11, Elsevier Science Publishers B.V., Amsterdam.
60. Koval, S.F. and Murray, R.G.E. 1986. Microbiol. Sci. 3, 357-361.
61. Smit, J. 1987. in Bacterial Outer Membranes as Model Systems. Inouye, M., Ed., pp. 343-376, John Wiley and Sons Ltd., New York.
62. Sleytr, U.B. and Messner, P. 1988. J. Bacteriol. 170, 2891-2897.
63. Ishiguro, E.E., Kay, W.W., Ainsworth, T., Chamberlain, J.B., Austen, R.A., Buckley, J.T. and Trust, T.J. 1981. J. Bacteriol. 148, 333-340.
64. Munn, C.B., Ishiguro, E.E., Kay, W.W. and Trust, T.J. 1982. Infect. Immun. 36, 1069-1075.
65. Kay, W.W., Phipps, B.M., Ishiguro, E.E. and Trust, T.J. 1985. J. Bacteriol. 164, 1332-1336.

66. Trust, T.J., Kay, W.W. and Ishiguro, E.E. 1983. Curr. Microbiol. 9, 315-318.
67. Owen, P. 1981. in Organization of Prokaryotic Cell Membranes, Vol.1, Ghosh, B.K., Ed., pp. 73-164, CRC Press, Inc., Boca Raton, Florida.
68. Owen, P. 1983. in Electroimmunochemical Analysis of Membrane Proteins Bjerrum, O.J., Ed., pp. 55-76, Elsevier Science Publishers, Amsterdam.
69. Owen, P. 1983. in Electroimmunochemical Analysis of Membrane Proteins Bjerrum, O.J., Ed., pp. 347-373, Elsevier Science Publishers, Amsterdam.
70. Owen, P. and Kaback, H.R. 1978. Proc. Natl. Acad. Sci. USA. 75, 3148-3152.
71. Owen, P. and Kaback, H.R. 1979. Biochemistry 18, 1413-1422.
72. Owen, P. and Kaback, H.R. 1979. Biochemistry 18, 1422-1426.
73. Owen, P., Kaczorowski, G.J. and Kaback, H.R. 1980. Biochemistry 19, 596-600.
74. Owen, P., Kaback, H.R. and Graeme-Cook, K.A. 1980. FEMS Microbiol. Lett. 7, 345-348.
75. Condon, C. and Owen, P. 1982. FEMS Microbiol. Lett. 14, 217-221.
76. Owen, P. 1986. Electrophoresis 7, 19-28.
77. Doherty, H., Yamada, H., Caffrey, P. and Owen, P. 1986. J. Bacteriol. 166, 1072-1082.
78. Ichihara, S. and Mizushima, S. 1979. Eur. J. Biochem. 100, 321-328.
79. McDade Jr.. R.L. and Johnston K.H. 1980. J. Bacteriol. 141, 1183-1191.
80. Newhall, W.J., Sawyer, W.D. and Haak, R.A. 1980. Infect. Immun. 28, 785-791.
81. Swanson, J. 1981. Infect. Immun. 34, 804-816.
82. Owen, P., Caffrey, P. and Josefsson, L.-G. 1987. J. Bacteriol. 169, 3370-3777.
83. Caffrey, P. and Owen, P. 1989. J. Bacteriol. 171, 3634-3640.
84. Korhonen, T.K., Rhen, M., Vaisanen-Rhen, V. and Pere, A. 1985. in Immunology of the Bacterial Cell Envelope Stewart- Tull, D.E.S. and Davies, M., Eds., pp. 319-354, John Wiley and Sons Ltd., Chichester, U.K.
85. Klemm, P. 1984. Eur. J. Biochem. 143, 395-399.
86. Klemm, P. 1985. Rev. Infect. Dis. 7, 321-340.
87. Sleytr, U.B. and Messner, P. 1983. Annu. Rev. Microbiol. 37, 311-339.
88. Hinson, G., Knutton, S., Lam-Po-Tang, M. K.-L., McNeish, A.S. and Williams, P.H. 1987. Infect. Immun. 55, 393-402.
89. Abraham, J.M., Freitag, C.S., Clements, J.R. and Eisenstein, B.I., 1985. Proc. Natl. Acad. Sci. USA 82, 5724-5727.
90. Klemm, P. 1986. EMBO J. 5, 1389-1393.
91. Smit, J., Grano, D.A., Glaeser, R.M. and Agabian, N. 1981. J. Bacteriol. 146, 1135-1150.
92. Charbit, A., Boulain, J.C., Ryter, A. and Hofnung, M. 1986. EMBO J. 5, 3029-3037.

93. Freudl, R., MacIntyre, S., Degen, M. and Henning, U. 1986. J. Mol. Biol. 188, 491-494.
94. van Die, I., Riegman, N., Hoekstra, W. and Bergmans, H. 1989. this volume.

95. Klemm, P. 1981. Eur. J. Biochem. 117, 617-627.
96. Fader, R.C., Duffy, L.K., Davis, C.P. and Kurosky, A. 1982. J. Biol. Chem. 257, 3301-3305.
97. Waalen, K., Sletten, K., Froholm, L.O., Vaisanen, V. and Korhonen, T.K. 1983.. FEMS Microbiol. Lett. 16, 149-151.
98. Schmoll, T., Hacker, J. and Goebel, W. 1987. FEMS Microbiol. Lett. 41, 229-235.

III. EVASION OF HOST DEFENSES

STRUCTURAL ASPECTS OF LPS: ROLE IN EVASION OF HOST DEFENSE MECHANISM

N. Grossman[1], A.A. Lindberg[2], S.B. Svenson[3], K.A. Joiner[4], and L. Leive[4]

[1]Department of Microbiology, The George S. Wise Faculty of Life Sciences, Tel Aviv University, Tel-Aviv, 69978, Israel; [2]Karolinska Institute, Huddinge Sweden; [3]National Biological Laboratory, Stockholm Sweden; and [4]NIH, Bethesda, MD, USA

INTRODUCTION

Lipopolysaccharide (LPS) is an important virulence determinant of gram-negative bacteria. It is comprised of three regions: the O-antigenic polysaccharide chain, the R-specific core poly- saccharide, and the lipid A region, which exerts, when soluble, all the endotoxic activities of LPS.

Interaction of LPS, specifically, the O-antigenic poly-saccharide moiety, with complement component C3 is critical in determining the ability of bacteria to evade two important host defense mechanisms: phagocytosis and complement-mediated. The experiments summarized in this article prove that the magnitude of this interaction is affected by two structural properties of the O-antigen in LPS: (a) The primary structure of the O-anti-genic repeating subunit, which varies within bacterial species and forms the basis for O-specific sero-typing; and (b) The length distribution of the polymerized O-specific polysaccharide chains, which varies to contain from 0 to over 100 O- antigenic repeating subunits (1,2).

PRIMARY STRUCTURE OF O-ANTIGEN AS A VIRULENCE DETERMINANT

To analyze the interaction of O-antigen in bacterial LPS with C3 and the complement cascade, *Salmonella* strains carry-ing three different well-characterized O-specific types of LPS, belonging to serogroups B (O-4,12), D (O-9,12) and C1 (O-6,7) (3,3,4, respectively) were compared.

Virulence of these strains varies according to the structure of their O-antigen (5,6,7,8) as expressed by LD50 values measured after i.p. injection to mice (Table 1): The strain carrying the O-4,12 serotype is the most virulent, the strain carrying the O-9,12 is of intermediate virulence, and the strain carrying the O-6,7 serotype was the least viru-lent. Similar results were obtained with immunosuppressed mice (9), suggesting that the differences in virulence did

Microbial Surface Components and Toxins in Relation to Pathogenesis
Edited by E.Z. Ron and S. Rottem, Plenum Press, New York, 1991

143

Table 1. Effect of O-Antigen Structure On Parameters Of
Bacterial Virulence

Type of LPS	Structure of O-Ag[a]	LD50 After i.p. Injection in Mice[b]	Rate of Uptake by Macrophages[c]	Percent of C3 Activation[d]
O-4,12	Abe (Man-Rha-Gal)n	2×10^4	9	18.1
O-9,12	Tyv (Man-Rha-Gal)n	5×10^4	29	43.5
O-6,7	Glc (Man-Man-Man-Man-GlcNAc)n	1×10^6	51	50.7

Man, Mannose; Rha, rhamnose; Gal, galactose; Glc, glucose;
GlcNAc, N- acetylglucoseamine; Abe, abequose; Tyv, Tyvelose.
[b] from refs 5,6.
[c] Uptake of bacteria/macrophage per 4 hr, from ref 11.
[d] Percent C3 removed during 1 hr incubation of bacteria
$(2 \times 10^8$ cells/ml) in absorbed C4D GPS (2%) relative to control serum, from ref 13.

not result from differential antibody induction. The differences in virulence were correlated with differences in clearance rate, suggesting that phagocytosis might be involved
(10).

EXPERIMENTS WITH INTACT BACTERIAL CELLS

    To analyze this differential virulence, an *in vitro*
assay system for phagocytosis was established in collaboration with H. Makela and colleagues (11). Three *Salmonella*
derivatives, constructed genetically to differ only in their
O-antigenic side chain and carrying the structures characteristic of serogroups B, D, and C1 (5,6), were incubated with
mouse macrophages in fetal bovine serum at 37°C. The bacterial cells were taken up by macrophages, and the rate and
extent of uptake of each strain were inversely proportional
to its virulence (11, Table 1): The strain with O-6,7 serotype, the least virulent, was taken up the best; the strain
with O-4,12 serotype, the most virulent, was taken up the
least, and the strain with O-9,12 serotype which was intermediate in virulence, was taken up at an intermediate rate.

    Bacterial uptake was not affected by changing the serum
source, or by the depletion of antibodies or fibronectin from
the sera. Only treatments that depleted or inactivated complement component C3, destroyed differential phagocytosis.
Therefore, we concluded (a) that the bacteria were taken up
by macrophages via complement receptors, and (b) that these
strains differ in their interaction with C3 (11).

    The three strains activated C3 at different rates and
extent, proportional to the extent of their phagocytosis (12,

Table 1). (Activation was measured either by measuring bound radiolabeled C3b, or by assaying remaining C3 after incubation in C4-deficient guinea pig serum, C4D GPS). The most virulent strain activated C3 the least, the least virulent strain activated C3 the best. This activation was initiated via the alternative pathway and did not require specific antibodies.

EXPERIMENTS WITH MODEL PARTICLES COATED WITH LPS DERIVATIVES

To eliminate the effect of bacterial cell surface components other than LPS on this differential complement activation, model particles were prepared: sheep erythrocytes, which do not activate complement via the alternative pathway, were coated with base-hydrolysed LPS, purified from the three bacterial strains. These preparations mimicked whole cells and activated C3 differentially, proving that LPS alone might be the cause for this phenomenon (13, Table 2).

In collaboration with Lindberg and Svenson we used a more purified system based on model molecules - oligosaccharide lipid conjugates (OLC) - containing O- specific oligosaccharides, but not lipid A or the core oligosaccharide moieties of LPS. Oligosaccharides, as short as two O-antigenic subunits, were prepared from base- hydrolysed LPS using phage-generated endoglucanases, and were then conjugated to octyl residues (14-17). Such conjugates generated from the three strains, were coated onto sheep erythrocytes and were tested for their ability to activate complement, when incubated in C4D GPS. These preparations activated C3 differentially (Table 2), and proportionally to particles coated with purified LPS, although the structure of the phage-generated O-specific subunit was slightly different than the biological subunit (15, Table 2).

These results prove that C3 activation depends exclusively on the structure of the O-antigen structure. Some variations in structure, such as the difference in the position of the hydroxyl group between antigen 4 (abequose) and antigen 9 (Tyvelose) result in a significant difference in C activation. On the other hand, a modification such as insertion of factor 5, O-acetyl (18), or the difference between the biological O-antigen subunit to the phage- generated subunit in the position of the O-substituting saccharide (15) has almost no effect on C3 activation.

IN VIVO STUDIES

Two years ago Helen Makela and coworkers showed that the *in vivo* situation was similar to this *in vitro* model (19). They have looked at the fate of radiolabeled live cells of O-6,7 strain and O-4,12 strain in the peritoneal cavity of mice. O-6,7 cells, the least virulent, were killed more extensively than the more virulent O-4,12 cells by resident peritoneal macrophages. They showed electron microscopically that bacterial cells were associated with macrophages and that they were coated with electron dense material that reacted with anti-C3, and was missing upon incubation in heat inactivated serum.

Table 2. Differential C3 Activation By Sheep Erythrocytes Coated With LPS derivatives.

Type of LPS	C3 Activation[a] by E-LPS[b]	Structure of OLC[c]	C3 Activation[a] by E-OLC[d]
O-4,12	0.05	Abe (Gal-Man-Rha)$_2$	0.05
O-9,12	0.25	Tyv (Gal-Man-Rha)$_2$	0.35
O-6,7	0.50	(Man-GlcNAc-Man-Man-Man)$_2$	0.52

[a] C3 activation is expressed in C3 units removed from serum, after 1 hr incubation of coated erythrocytes in 2% C4D GPS, per LPS or OLC molecule ($X10^{-5}$) coated on erythrocytes.
[b] Erythrocytes coated with base-hydrolysed LPS.
[c] Man,Mannose; Rha, rhamnose; Gal, galactose; GlcNAc, N-acetylglucoseamine; Abe, abequose; Tyv, Tyvelose.
[b] Erythrocytes coated with oligosaccharide-lipid conjugates (OLC).

HOW IS THIS DIFFERENTIAL ACTIVATION REGULATED?

Two complement components compete after the first C3b-like molecule is deposited on a cell surface: factor B and factor H. If <u>factor B</u> has a greater affinity, activation continues and more C3b molecules are deposited on the activating surface. If the second factor, <u>factor H</u>, has a greater affinity to the surface-bound C3b, it cleaves the C3b molecule to produce iC3b, this way abolishing further activation and deposition of more C3b molecules on this surface. In addition, this cleavage block build-up of the membrane attack complex (20).

Recently, Jimenez-Lucho et al. (21) showed that there may be two mechanisms responsible for the differences among the strains: The O-4,12 strain binds factor B less than the other two strains, hence making a poorer C3 convertase; The O-9,12 binds less C3b molecules initially than the other two strains. The three strains however bind factor H with the same affinity. Consequently, a similar fraction of C3b molecules is converted to C3bi molecules (22).

THE ROLE OF O-ANTIGEN LENGTH IN EVASION OF COMPLEMENT MEDIATED LYSIS

Purified LPS can be demonstrated on SDS-PAGE to be a population of molecules varying in their O-antigen chain length, with a characteristic size distribution (1,2). Such size distributions can be quantitated when advanced densitometry methods are applied to analysis of radiolabeled LPS (1,23,24).

Cells opsonized with C3b get lysed by insertion of

membrane attack complexes comprised of C5b-9 into their membrane. It appears that the length of the O-polysaccharide side chain affects the extent of killing. Accordingly, cells bearing the complete set of molecules are known to be serum resistant (i.e. do not get lysed unless antibodies are added). This resistance was shown to result from weak binding of C9 to C5b-8 complexes on smooth *Salmonella minnesota* and was then released upon further incubation (25). A similar observation was obtained for *E. coli* 0111 by Goldman, Joiner and Leive (26).

THE EXPERIMENTAL SYSTEM

To be able to vary the length and/or the size distribution of LPS molecules we have used a mutant strain of *S. montevideo*, constructed by H. Makela and B. Stocker (23), that requires exogenous galactose and mannose to build its core oligosaccharide and its O-antigen, respectively.

The pattern of LPS synthesized by this strain was modulated by growing the cells with enough galactose but with varying concentrations of mannose (23). When no mannose was supplied, LPS molecules contained only Lipid A and core oligosaccharide. When the cells were supplemented with optimal concentration of mannose, the complete set of long chain LPS was synthesized, i.e. about 30% of total LPS molecules were longer than 14 O-repeating units/LPS molecule. When the cells were grown with suboptimal concentrations of mannose, two changes were observed: the average length of the O-antigenic polysaccharide chains became shorter, and the fraction of molecules with long O-antigenic chain decreased with decreasing mannose concentration. Yet, the fraction of LPS molecules that contained O-antigen was not affected by growth conditions, suggesting indicating an advantage to the cells.

Lysis of cells incubated in serum, occurred only if they were pregrown with suboptimal concentration of mannose (23). We concluded that under conditions that allowed a large fraction total LPS molecules to contain long O-antigen chains was sterically hindering the membrane attack complex from insertion into its target site, i.e. the bacteria became protected against complement-mediated killing. In other systems, *E. coli* 0111 (26), and *E. coli* O6 and O18 (27), protection was bacteria from killing was associated with a larger fraction of O-antigen bearing LPS molecules, in addition to increased length of the O-antigenic polysaccharide.

It is important to mention that although C3 bound preferentially to the longest LPS molecules available under all mannose concentrations employed (28), the length of O- antigenic polysaccharide chain probably plays a role in evading complement-mediated lysis and not in modulating complement activation. This conclusion is based on experiments where erythrocytes coated at similar densities with base-hydrolysed LPS, with an average of 10 O-antigenic repeating units (13), activated C3 similarly to erythrocytes coated with OLC (two O-antigenic subunits, Table 2) and to erythrocytes coated with size-fractionated LPS, with an average of three O-antigenic subunits (13).

CONCLUSIONS

The structural properties of LPS are involved at least in three mechanisms used by bacteria to evade host defense:
(a) The primary structure of O-antigen determines evasion of phagocytosis by determining the extent of complement activation via the alternative pathway;
(b) The primary structure affects the fate of C3b associated with the activating surface, thus determining whether formation of membrane attack complex is proceeded;
(c) The length distribution of O-antigen side chains determines whether this complex is completed, and if the complex is formed - it determines evasion of complement mediated lysis by hindering access to the cell membrane.

These studies demonstrate general strategies that microorganisms other than *Salmonellae* may follow while evading host defenses. Accordingly, strains that undergo phase variation are likely to evade host defense due to a change in the primary structure of surface components, while encapsulated bacteria may be protected due to steric hindrance.

ACKNOWLEDGEMENTS

This article is dedicated to the memory of Loretta Leive (deceased February, 12, 1986), an excellent scientist, colleague and friend. Special thanks are awarded to Helen Makela, our long-term collaborator, for her continuous support and friendship.

REFERENCES

1. Goldman, R.C. and L. Leive. 1980. Eur. J. Biochem. 107: 145-153.
2. Palva, T.E., and P.H. Makela. 1980. Eur. J. Biochem. 107:137-144.
3. Nurminen, M., Hellerequist, C.G., Valtonen, V.V., and P.H. Makela. 1971. Eur. J. Biochem. 22: 500-505.
4. Fuller, N.A., and A.M. Staub. 1968. Eur. J. Biochem. 2: 286-300.
5. Valtonen M.V., Plosila, M., Valtonen, V.V. and P.H. Makela. 1975. Infec.Immun. 12: 828-835.
6. Valtonen, V.V.. 1970. J. Gen. Microbiol. 64: 255-268.
7. Valtonen, V.V., Aird, J., Valtonen, M., Makela, O., and P.H. Makela. 1971. Acta Path. Microbiol. Scand. 79B, 710-715.
8. Valtonen, M.V. and P. Hayry. 1978. Infec. Immun. 19: 26-28.
9. Makela, P.H., Valtonen, V.V., and M.V. Valtonen. 1973. J. Infec. Dis. 128. Suppl..: S81-S85.
10. Valtonen M.V.. 1977. Infec. Immun. 18: 574-578.
11. Liang-Takasaki, C.-J., Makela, P.H. and L. Leive. 1983.. J. Immunol. 128: 1229-1235.
12. Liang-Takasaki, C.-J., Grossman, N., and L. Leive. 1983. J. Immunol. 130: 1867-1870.
13. Grossman, N., and L. Leive. 1984. J. Immunol. 132: 376-385.
14. Svenson, S.B., and A.A. Lindberg. 1977. FEMS Microbiol. Lett. 1: 145-148.

15. Svenson, S.B., and A.A. Lindberg. 1978. J. Immunol. 120: 1750-1757.
16. Svenson, S.B., and A.A. Lindberg. 1979. J. Immunol. Methods 25: 323-335.
17. Robertsson, J.A., Svenson, S.B., and A.A. Lindberg (1982. Infec. Immun. 37: 737-748.
18. Valtonen, V.V. and P.H. Makela. 1971. J. Gen. Microbiol. 69: 107-115.
19. Makela, P.H., Hovi, M., Saxen, H., Valtonen, M., and V. Valtonen. 1988. Immunol. Lett. 19: 217-222.
20. Joiner, K.A.. 1988. Ann. Rev. Microbiol. 42: 201-230.
21. Jimenez-Lucho, V.E., Joiner, K.A., Foulds, J., Frank, M.M., and L. Leive. 1987. J. Immunol. 139: 1253-1259.
22. Grossman, N., Joiner, K.A., Frank, M.M., and L. Leive (1986. J. Immunol.136: 2208-2215.
23. Grossman, N., Schmetz, M., Foulds, J., Klima, E.N., Jiminez, V., Leive, L., and K.A. Joiner.. 1987. 169: 856-863.
24. Pun, T., Trus, B.L., Grossman, N., Leive, L., and M. Eden. 1985. Electrophoresis 6: 268-273.
25. Joiner, K.A., Hammer, C.H., Brown, E.J., and M.M. Frank (1982. J. Exp. Med. 155: 797-815.
26. Goldman, R.C., Joiner, K.A., and L. Leive. 1984. J. Bacteriol. 159: 877-882.
27. Porat, R., Johns, M.A., and W.R. McCabe. 1987. Infec. Immun. 55: 320-328.
28. Joiner, K.A., Grossman, N., Schmetz, M., and L. Leive (1986. J. Immunol. 136: 710-715.

CHEMICAL CHARACTERIZATION OF *E. coli*

CAPSULES AND ANALYSIS OF THEIR EXPRESSION

Barbara Jann, Maria-Luisa Rodriguez, Andreas Finke,
Klaus-Dieter Kröncke and Klaus Jann

Max-Plank-Institut für Immunbiologie, Freiburg, FRG

INTRODUCTION

Escherichia coli causing extraintestinal infections
counteract the unspecific host defense with the formation of
capsules. These consist of acidic polysaccharides which form
a gel-like extracellular structure with a very high water
content. Fig. 1 shows an encapsulated *E. coli* after incubation with specific anticapsular antibodies. Due to their
exquisite structure, capsules collapse during preparation of
encapsulated bacteria for electron microscopy, unless they
are stabilized. Today some 70 serologically distinct capsular
polysaccharides are known which form capsules of similar
appearance. The primary structures of about 40 of these have
been determined (1). They consist of repeating oligosaccharide units and their structures determine the serological K
specificity of encapsulated bacteria. Modern techniques, such
as NMR spectrometry and computer-aided molecular modelling
permits the analysis of secondary structures and thus the
presentation of epitopes.

GENERAL CHARACTERIZATION AND STRUCTURE

On the basis of several parameters, we have subdivided
the capsular polysaccharides into two groups (Table 1) (1,2).
The principal differentiating features are: molecular weight,
nature of the acidic component, temperature regulation of
expression, occurrence in different O groups (i.e. co-expression with O antigenic lipopolysaccharides), and genetic
determination. Coli bacteria with group II polysaccharides
are of special interest in extraintestinal coli infections.
These polysacchrides occur in many O groups and they are not
expressed at lower growth temperatures (18-20°C) (3). They
are substituted at the reducing end with a phosphatidic acid
in a rather labile linkage. Interestingly, group II capsular
polysaccharides are similar to the capsular polysacchrides of
N. meningitidis and *H. influenzae,* not only in structure but
also with respect to pathogenicity of the respective bacteria. A number of group II polysacchrides contain 2-keto-3--
deoxy-$\underline{D}$ -manno octonic acid (KDO) in their repeating unit.

Microbial Surface Components and Toxins in Relation to Pathogenesis
Edited by E.Z. Ron and S. Rottem, Plenum Press, New York, 1991

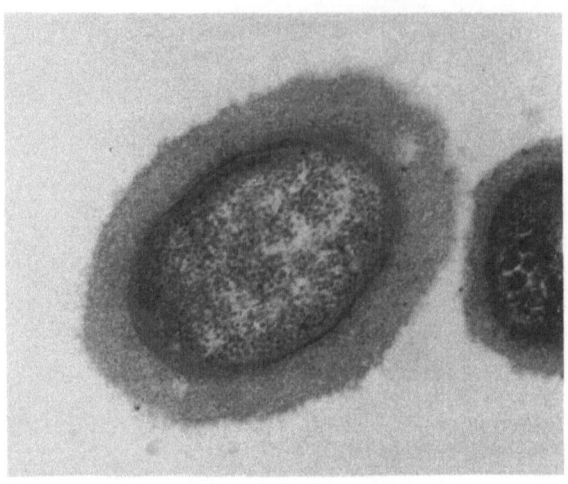

Fig. 1.   Electronmicroscopic picture of a thin section from
           E. coli O25:K19. The capsule was stabilized with
           homologous antibody.

Fig. 2 shows the structure of KDO and its structural and
biochemical relation to N-acetylneuraminic acid. Most KDO
containing capsular polysaccharides consist of disaccharide
repeating units (Table 2), six out of eight having β-L-ribose
as the second constituent. From the four KDO containing polysaccharides with a trisaccharide repeating unit (Table 3)
three also contain only ribose as additional constituent
(1,4). Many of the KDO containing polysaccharides are structurally very closely related. It will be interesting to
compare their secondary structures in order to elucidate the
differences of epitope presentation in this family of capsular polysaccharides.

Table 1.   A classification of *E. coli* capsular polysac-
               charides into group I (related to *Klebsiella* capsu-
               lar antigens) and group II (related to capsular
               antigens of *N. meningitidis* and H. influenzae).

Properties	I	II
molecular weight	>100,000	<50,000
acidic component	hexuronic acid	(glucuronic acid)
	pyruvate	phosphate
		KDO
		NeuNAc
stability at pH 6, 100°C	all stable	most stable
co-expressed with	O8, O9, (O20)	many O antigens
chromosomal determination	<u>his</u>	<u>serA</u>
expressed at 17-20°C	yes	no

Table 2. KDO containing capsular polysaccharides with disaccharide repeating units.

K antigen	repeating unit	KDO linkage
K19	3-ßRib-1,4-ßKDO-2, *	4-KDOp
K97	2-ßRib-1,5-ßKDO-2,	
K14	6-ßGalNAc-1,5-ßKDO-2, *	5-KDOp
K15	4-ßGlcNAc-1,5-ßKDO-2,	
K13	3-ßRib-1,7-ßKDO-2, *	
K20	3-ßRib-1,7-ßKDO-2, *	7-KDOp
K23	3-ßRib-1,7-ßKDO-2,	
K95	3-ßRib-1,8-ßKDO-2, *	8-KDOf

*Asterisks indicates acetylation.

Table 3. KDO containing capsular polysaccharides with trisaccharide repeating units.

K antigen	repeating unit	KDO linkage
K12	3-αRha-1,2-αRha-1,5-ßKDOp-2, *	5-KDOp
K16	2-ßRib-1,3-ßRib-1,5IßKDOp-2, *	5-KDOp
K6	2-ßRib-1,2-ßRib-1,7-αKDOp-2, *	7-KDOp
K74	3-ßRib-1,2-ßRib-1,6-ßKDOf-1, *	6-KDOf

*Asterisks indicates acetylation.

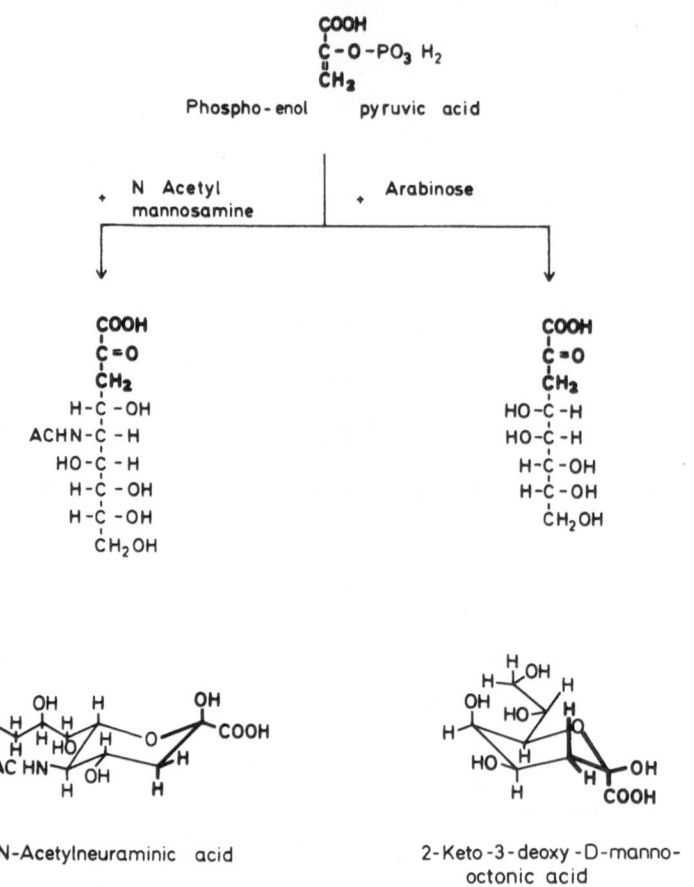

Fig. 2. Biochemical relationship between N-acetylneuraminic acid (NeuNac) and 2-keto-3-deoxy-D-manno-octonic acid (KDO).

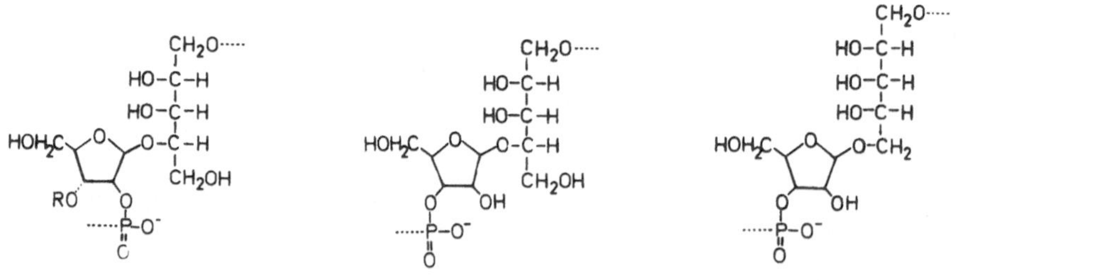

Fig. 3. Repeating units of the K22 (1), K18 (2), K100 (3) antigens of *E. coli* and of the capsular antigens of *H. influenzae* b (4).

H. INFLUENZAE b CPS   E. COLI K100 CPS

TOP VIEW   TOP VIEW

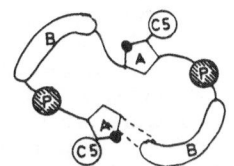   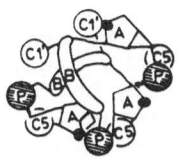

SIDE VIEW   SIDE VIEW

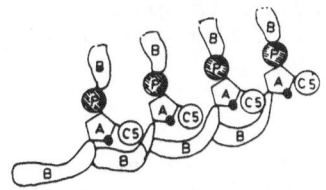   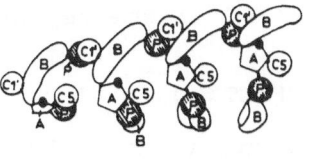

A: Ribose   C1': position of ribitol   P. phosphate of
B: Ribitol   C5 : position of ribitol   phosphodiester

Fig. 4.   Schematic drawings of the secondary structure of
the capsular antigen from *Haemophilus influenzae*
type b and *E. coli* K100 as obtained by computer
aided modelling.

    We have recently undertaken such an analysis of second-
ary structures with the *E. coli* K18, K22 and K100 capsular
antigens and the capsular antigen of *H. influenzae* b (Hib).
These antigens are related ribosyl-ribitol phosphates (Fig.
3) (5). In a collaboration with A. Neszmelyi/Budapest we
could show with two dimensional NMR spectrometry and with
computer aided molecular modelling that the capsular K18,
K22, K100 and Hib capsular antigens form helices with different pitches and different surface expression of components.
In Fig. 4, the helices of the K100 and the Hib polymers are
compared. In the helix formed by the Hib capsular antigen the
repeating units on subsequent turns are superimposed and the
pitch as well as the oval cross section of one plane allows
the presentation of all substructures. In contrast, the subunits on subsequent turns of the K100 helix are twisted (skew
positions) and the cross sections are smaller, due to the
different linkage position in the ribitol unit. This results
in the exposure of the ribosyl-ribitol part of the repeating
unit and in a non-exposed position of the ribitol-phosphate
part. Thus, the latter is not cross reactive between K100 and
Hib antigens. An important feature of these polysaccharides

seems to be the relative positions of charges alongside the helix, and this is different between the K22, K100 and Hib antigens (6).

One of the problems in capsule research is the question of how these structures are retained by the cell in spite of their hydrophilicity and high charge. It was found (as shown in Fig. 8) that the group II capsular polysaccharides exhibited a slow moving band and a fast moving band in immunoelectrophoresis. Further, gel chromatography indicated the presence of a high molecular weight and a low molecular weight fraction in the polysaccharide preparations (7). After mild acid treatment or in the presence of detergent the high molecular weight band disappeared and the polysaccharides showed only the fast moving band in immunoelectrophoresis. Mild acid hydrolysis removed a phosphatidic acid from the reducing end of the respective polysaccharide. The structure of lipid substituted capsular polysaccharide is shown in Fig. 5 with the K12 antigen as an example. In group II polysaccharides with KDO in the repeating unit the reducing end is KDO.

ROLE IN INFECTION

It is known for a long time that encapsulation renders pathogenic bacteria resistant, or at least less vulnerable to the host defense. This has been shown for serum resistance as well as for resistance to phagocytosis (8). It is not always the capsule alone which exerts this protection but rather its combination with certain O antigenic lipopolysaccharides, e.g. the K1 antigen with the O1 or O18 antigen of the K5 poly -saccharide with the O18 or the O75 antigen (9). The protective effect of a capsule is normally overcome by specific anticapsular antibodies which are produced later in the infection. However, in some instances a structural identity or similarity of the capsular polysaccharide with complex carbohydrates of the host prevent or drastically reduce immune recognition and thus immunogenicity of the capsular polysaccharide. As a result antibodies against these capsular polysaccharides are not or only very poorly formed, and bacteria disguised with such non-immunogenic capsules are very virulent. Fig. 6 shows the structures of the K1 polysaccharide, which has the same structure as the carbohydrate

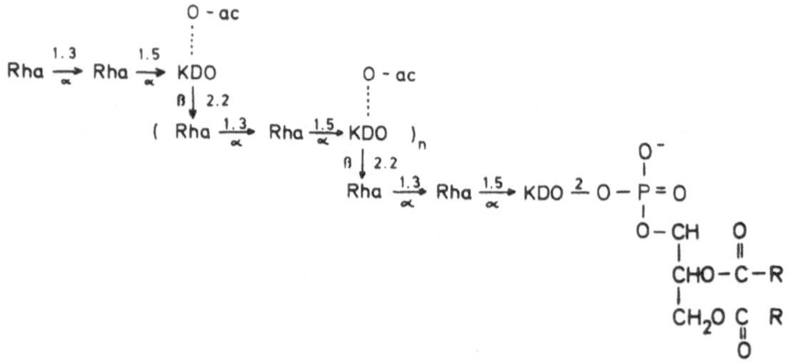

Fig. 5.   Structure of the K12 antigen of *E. coli* with the terminal substitution of phosphatidic acid.

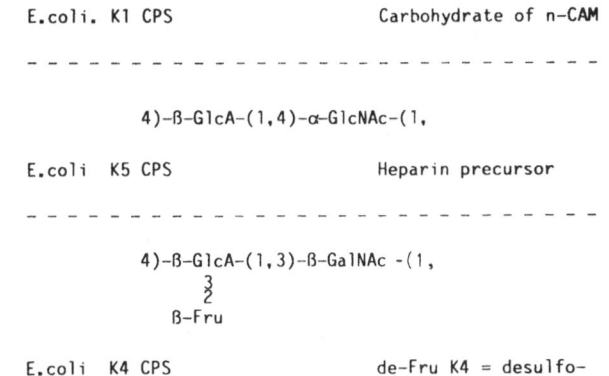

8)-α-NeuNAc-(2,

E.coli. K1 CPS   Carbohydrate of n-CAM

- -

4)-β-GlcA-(1,4)-α-GlcNAc-(1,

E.coli K5 CPS   Heparin precursor

- -

4)-β-GlcA-(1,3)-β-GalNAc -(1,
3
2
β-Fru

E.coli K4 CPS   de-Fru K4 = desulfo-
chondroitin

Fig. 6.   Structure of the K1, K5 and K4 capsular polysaccharides and their relation to mammalian complex carbohydrates.

terminus of the neural cell adhesion molecule nCAM of developing brain (10) and the structure of the K5 polysaccharide,
which has the same structure as the first polymeric intermediate of heparin (11). Fig. 6 also shows the structure of the
K4 polysaccharide which we have recently elucidated (12).
This fructose containing polysaccharide has the same backbone
as chondroitin sulfate. Removal of fructose from the K4
polysaccharide occurs very easily at low pH and the defructo-
sylated polysaccharide has lost its K4 specificity. It is
possible that in body compartments with low pH E. coli K4
bacteria lose fructose from the capsule and are then covered
with a non immunogenic capsule.

Interestingly, capsules seem at least in some cases, to
be penetrable for antibodies against the underlying O antigenic lipopolysaccharide. Fig. 7 shows the opsonophagocytosis
of E. coli O4:K12 with monoclonal anti O4, anti-K12 antibodies and with both (13). It is obvious that killing of the
encapsulated bacteria by phagocytes occurs to a large extent
after opsonization with the anti-O antibody. We have subjected these bacteria, which were all encapsulated, to an immuno-
electronmicroscopic analysis. It was shown that in whole cell
preparations the cell wall lipopolysaccharide can be labelled
in spite of the capsule. An electronmicrograph of a thin
section, in which the capsule was contrasted with anti-K
antibody and the lipopolysaccharide was labelled with the
immunogold technique revealed that the capsule contains
patches of lipopolysaccharide (14). These may well be accessible from outside.

BIOSYNTHESIS AND SURFACE EXPRESSION

It was found that the genes determining the biosynthesis
and expression of group II capsular polysaccharides are
organized in three regions (15). A central region which
determines the primary structure of the capsular polysaccha-

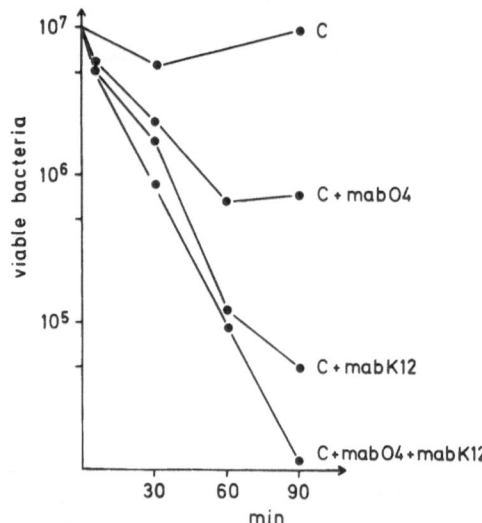

Fig. 7.   Phagocytosis of human PMN of *E. coli* O4:K12 opsonized with   complement (C) in the   absence or presence of monoclonal antibodies (mab).

ride is flanked by two regions determining translocation and
surface expression. Whereas the polymerization genes in the
central region are characteristic of *E. coli* with serologically distinct capsules, the flanking regions have a common
structure and function in *E. coli* with group II capsular
polysaccharides (16).

    The polymerization directed by the central gene region
was analyzed with the K5 polysaccharide as an example. We
found that no lipid co-factor is needed for chain elongation
and that the polysaccharide chain grows at the non reducing
end by sequential addition of the component sugars from their
nucleotide precursors.

    The reducing end of the K5 polysaccharide was found to
be KDO (Finke, *et al.*, in preparation). Since KDO is not a
constituent of the repeating unit, this finding may be of
consequence. Under the assumption that transfer of KDO from
its activated form CMP- KDO to an acceptor may initiate the
biosynthesis of the K5 polysaccharide, we analyzed this
enzyme activity in cytoplasmic fractions. Compared to un-
capsulated and rough strains, this enzyme activity was significantly elevated in *E. coli* K5 (17). We could correlate
the elevated CMP-KDO synthetase activity with the biosynthesis of the K5 polysaccharide. An analysis of K5 clones, as
well as insertion and deletion mutants indicated that the
gene(s) determining elevated CMP-KDO synthetase activity are
in region 1, one of the general gene regions determining
expression. Subsequent analysis of many coli strains with
group II polysaccharides, including K1, showed that they had
an elevated CMP-KDO synthetase activity (17a). From this we
postulate as a working hypothesis that the biosynthesis of
group II polysaccharides generally starts with the transfer
of KDO from CMP-KDO to a membrane-associated acceptor.

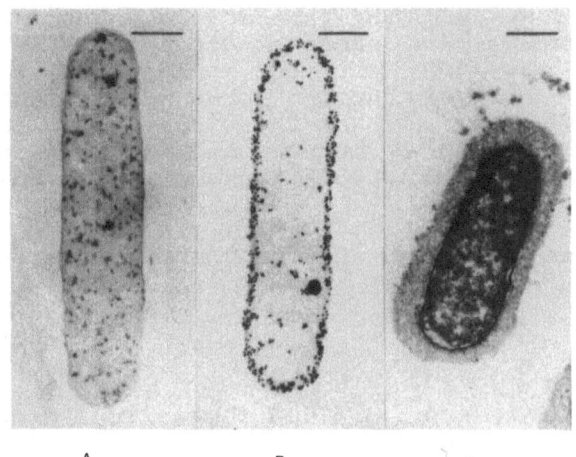

A   B   C

Fig. 8.   Electronmicrographs of ultrathin sections from
          region 3 mutant *E. coli* 21500 (A), region I mutant
          E. coli 21487 (B) ´ and wild type *E. coli* 21848 ex-
          pressing the K5 capsule (C). A and B were embedded
          in Lowicryl K4M and labelled with monoclonal anti-
          K5-antibodies and gold. The wild type (C) was em-
          bedded in Epon after incubation with the monoclonal
          anti-K5-antibodies.

     To study the surface expression of the K1, K5 and K12
capsular polysaccharides, we made use of the bacteria at that
capsule restrictive temperature, the suspension was warmed to
37°C and capsule expression was followed with the use of
capsule-specific bacteriophages and with anticapsular antibodies. Capsule expression began about 20 minutes after temperature upshift and was complete after another 25 minutes.
Immunoelectron microscopy showed that the expression started
not evenly over the entire cell surface but in clusters which
grew laterally (18). Plasmolysis of the bacteria prior to
capsule fixation and electron microscopy revealed that the
clusters were over sites where cytoplasmic and outer bacterial membranes come into contact.

     Further inside into the mechanism of surface transloca-
tion was obtained with *E. coli* K5 expression mutants (19). A
transport mutant defective in gene region 1 accumulated the
polysaccharide at the cell membrane (Fig. 8). After plasmolysis, the polysaccharide was labelled in the periplasm. Mutation in gene region 3, which was hitherto functionally ill
defined, lead to an accumulation of the capsular polysaccharide in the cytoplasm. This gene region, which encodes not
more that two proteins, seems to direct the translocation of
the polysaccharide across the cytoplasmic membrane and must
govern energy requiring processes.

     The polysaccharide isolated from the region 1 and region
3 mutants were isolated. The immunoelectrophoretic pattern of
the region 1 (transport) mutant had the same pattern as the
wild type polysaccharide. The preparation from the region 3

mutant exhibited only the fast moving band, indicating that it did not contain the phosphatidic acid. This was confirmed by chemical analysis. Polyacrylamide gel electrophoresis showed that the polysaccharide preparation from the region 1 mutant had the same size as the wild type polysaccharide and that from the region 3 mutant was smaller. This shows that polymerization and translocation must be correlated processes.

This work was supported by the Deutsche Forschungs-gemeinschaft (DFG).

REFERENCES

1.   Jann, B. and Jann, K. 1989. Curr. Top. Microbiol. Immunol. 150:19-42.
2.   Jann, K. and Jann, B. 1983. Prog. Allergy 33, 53-79.
3.   Orskov, F., Sharma, V. and Orskov, I. 1984. 130, 2681-- 2684.
4.   Jann, K. and Jann, B. 1988. NATA ASI Series, Vol. H24, 41- 53.
5.   Rodriguez, M.L., Jann, B., and Jann, K. 1988. Carbohydr. Res. 173:243-253.
6.   Rodriguez, M.L. 1988. Dissertation, University of Freiburg, FRG.
7.   Schmidt, M.A. and Jann, K. 1982. FEMS Microbiol. Lett. 14, 69-74.
8.   Kim, K.S., Kang, J.H., and Cross, A.S. 1986. FEMS Microbiol. Lett. 35, 275-278.
9.   Cross, A.S., Kim, K.S., Wright, D.C., Sadoff, J.C., and Gemski, P. 1986. J. Infect. Dis. 154, 497-503.
10.   Finne, J. 1985. Trends Biochem. Sci. 10, 129-132.
11.   Vann, W.F., Schmidt, M.A., Jann, B., and Jann, K. 1981. Eur. J. Biochem. 116, 359-364.
12.   Rodriguez, M.L., Jann, B., and Jann, K. 1988. Eur. J. Biochem. 177, 117-124.
13.   Abe, C., Schmitz, S., Jann, B., and Jann, K. 1988. FEMS Microbiol. Lett. 51:153-158.
14.   Kroncke, K.-D. 1989. Dissertation, University of Freiburg, FRG.
15.   Boulnois, G.J., Roberts, I.S., Hodge, R., Hardy, K.R., Jann, K., and Timmis, K.N. 1987. Mol. Gen. Genet. 209, 242-246.
16.   Roberts, I.S., Mountford, R., Hodge, R., Jann, K., and Boulnois, G.J. 1988. J. Bacteriol. 170, 1305-- 1310.
17.   Finke, A., Roberts, I.S., Boulnois, G., Pazzani, C., and Jann, K. 1989. J. Bacteriol. 171:3074-3079.
17a. Finke, A., Jann, B. and Jann, K. 1990. FEMS Microbiol. Lett. 69:129-134.
18.   Kroncke, K.D., Golecki, J., and Jann, K. 1989. J. Bacteriol. 172:3469-3472.
19.   Kroncke, K.D., Boulnois, G., Roberts, J., Bitter-Suermann, D., Golecki, J., Jann, B., and Jann, K. 1989. J. Bacteriol. 172:1085-1091.

STRUCTURE, FUNCTION AND ROLE IN DISEASE OF PNEUMOLYSIN,

THE THIOL-ACTIVATED TOXIN OF *STREPTOCOCCUS PNEUMONIAE*

G.J. Boulnois , T. Mitchell,
K. Saunders, X. Mendez and P. Andrew

Department of Microbiology, University of Leicester
Leicester, LE1 9HN, U.K.

INTRODUCTION

Pneumolysin, the thiol-activated toxin of *Streptococcus pneumoniae*, is one of a family of toxins produced by four different genera of Gram positive bacteria (1). This family of toxins share a variety of physical and biological properties and exert their effects via damage to eukaryotic membranes (2). A striking feature of this family is their pronounced immunological cross-reactivity such that sera raised against one member of this family generally reacts with and often neutralizes and precipitates heterologous toxin (1). They are termed thiol-activated since they are inactivated upon oxidation and treatment with reducing agents restores full activity (1). This was thought to reflect the formation and breakage of intra-molecular disulfide bridges, a process which induces conformational changes in the protein which are reflected in their ability to interact with membranes (3). As well as mediating such changes, a single sulphydryl was postulated to be essential for activity (4). However, the role this essential cysteine plays in toxin activity is unclear. The thiol-activated toxins are thought to utilize cholesterol as receptor since their cytolytic activity is only manifest on cells which have cholesterol as part of their membranes and since free cholesterol is a potent inhibitor of cytolytic activity (1,2). It has been postulated that this essential sulphydryl group may mediate (or is involved in) the interaction of the toxin and cholesterol (6). It should be noted that cholesterol has not been shown conclusively to act as the receptor for these toxins.

Once bound to target cells, the toxin is inserted into the lipid bilayer and following lateral diffusion form aggregates of toxin molecules (6) which are probably cholesterol-free (7). These aggregates are visible as arc and ring shape structures in the electron microscope and it is thought that they represent transmembrane channels through which cytoplasmic molecules can pass leading to cell lysis (8).

The relevance of these structures *in vivo* during infec-

Microbial Surface Components and Toxins in Relation to Pathogenesis
Edited by E.Z. Ron and S. Rottem, Plenum Press, New York, 1991

161

Pneumolysin	RECTGLAWEWWRTVY
SLO	RECTGLAWEWWRKVI
Perfringolysin	RECTGLAWEWWRDVI
Listeriolysin	KECTGLAWEWWRTVI

Fig. 1.   The cysteine motif of the thoil activated toxin. The unique cysteine residue in pneumolysin is residue 428.

tion is unclear since it has been argued that insufficient concentrations of these toxins are reached *in vivo* to permit their formation. However sublytic concentrations of toxin have potent inhibitory effects on cells of the immune system and such effects might have profound consequences for pathogenesis of infection caused by their producing organisms. For example, sub-lethal amounts of pneumolysin inhibit antibody synthesis by B cells (9) and inhibit the key antimicrobial activities of polymorphonuclear leukocytes (PMNL) such as the respiratory burst (10). Some of the thiol-activated toxins, including pneumolysin, activate the classical pathway of complement in an antibody independent fashion (11). When this occurs in the fluid phase this may consume complement thus abrogating its protective effects (7).

Toxin-membrane complexes are also potent activators of complement and this may result in assembly of the membrane attack complex on these membranes leading to autolysis (7,8). Each of these events may contribute to the induction of the inflammatory response.

To further probe the structure and function of pneumolysin and to study its role in the pathogenesis of pneumococcal infection we recently reported the cloning (12) and complete sequence of the pneumolysin gene (12) and its manipulation to yield large amounts of recombinant toxin (13). Subsequently the sequence of three other thiol-activated toxins, Streptolysin O (14), Listeriolysin (6) and Perfringolysin (15) have been deduced and in this short article we shall review progress to date drawing on the comparison of the available sequences and in particular our own studies on pneumolysin.

STRUCTURE OF THE THIOL-ACTIVATED TOXINS

Pneumolysin, unlike the other thiol-activated toxins, is not secreted from the producing organism (16). No signal sequence was found at the N-terminus of pneumolysin in contrast to the other three toxins so far analyzed (12). Pneumolysin is 471 amino acids in length and comparison of the sequence with the low molecular weight forms of Streptolysin O, and the secreted version of perfringolysin and listeriolysin reveals that they all share remarkable primary sequence similarities (6,12,14,15). This presumably accounts for their serological cross-reactivity. It is interesting to note that this similarity is in the context of only limited but detectable DNA sequence homologies of their respective genes (6, 14). This suggests that the genes for each of these toxins evolved from a common ancestor.

Table 1. Mutagenesis of the cysteine motif.

Modification	% activity	red cell binding	oligomer formation	cholesterol binding	complement activation
Cys$_{428}$ › ala	100	100	+	100	100
Cys$_{428}$ › ser	15	ND	+	100	100
Cys$_{428}$ › gly	1	100	+	100	100
Trp$_{433}$ › Phe	0.6	ND	+	100	ND
Glu$_{434}$ › Asp	25	ND	+	50	ND
Trp$_{435}$ › Phe	13	ND	ND	ND	ND
Trp$_{436}$ › Phe	100	ND	ND	ND	ND
Cys$_{428}$ › Gly ⌉ Trp$_{433}$ › Phe ⌋	0.1	ND	+/-	ND	ND
wildtype	100	100	+	100	100
recombinant	100	100	+	100	100

% activity refers to the hemolytic activity of the modified toxins relative to either wild-type recombinant or native pneumolysin. Hemolytic activities were determined by standard methods (ref.12). Red cell binding was measured by incubating cells with pneumolysin radiolabelled *in vitro* with ^{35}S and determining number of counts bound. Oligomer formation was determined by separation of toxin oligomers from red cell membranes using sucrose density centrifugation (ref.8) as described previously (ref.12). Cholesterol binding. Toxin was incubated with radiolabelled cholesterol and toxin-cholesterol complexes separated by sucrose density centrifugation. Binding was measured by counting label associated with toxin. Complement activation was measured by determination of C3b formation by immuno-electrophoresis as described previously (ref.17).

Each of these toxins contains a single cysteine residue towards the C terminus of the toxin molecule. This observation precludes the possibilities of intra-molecular disulfide bridge formation as the basis for reversible oxidation and reduction of these toxins.

The single cysteine is presumably the essential cysteine residue defined previously. Interestingly, the single cysteine residue lies in an 11 amino acid motif which represents the single largest region of homology between these toxins (Fig. 1). When the sequences of the individual toxins are aligned to give maximal similarity the single cysteine residues in each toxin are also aligned. The extensive homology of these toxins from three different genera presumably reflects conservation of overall structure required for activity.

Each of the thiol-activated toxins so far analyzed contains a common 11 amino acid sequence motif which contains the single cysteine and three tryptophan residues (6,12,14, 15). Together these observations point to this sequence motif being important for toxin activity. To explore the role of this motif in activity we have carried out an extensive analysis (Table 1) using oligonucleotide-mediated, site--directed mutagenesis of the pneumolysin genes to systematically change residues in this region (17, unpublished data).

The Cys_{428}>Ala modified toxin was indistinguishable from native and recombinant, wild-type toxin in all parameters so far examined, including cytolytic and sublethal effects on PMNLs and the ability to activate complement (17). This clearly precludes an essential role for cysteine residue and hence sulphydryl groups in pneumolysin activity *in vitro*. It remains possible that the sulphydryl group important *in vivo*. However, the nature of the amino acid at position 428 in pneumolysin is not unimportant since the Cys_{428}>Ser and Cys_{428}>Gly modified toxins had reduced cytolytic activity and a reduced ability to inhibit the respiratory burst of phorbol ester-stimulated human PMNL (17). The reasons for this are unclear but we cannot exclude disruption of the secondary structure of these toxins as an explanation for their reduced activity.

Modification of the tryptophan residues led to an interesting observation (Table 1, unpublished). The Trp_{433}>Phe and Trp_{435}Phe had reduced cytolytic activity whilst the Trp_{436}>Phe modified toxin was fully active. The Glu_{434}>Asp modified toxin was also reduced in cytolytic activity. Thus despite their pronounced conservation, specific residues (including the cysteine) are not absolutely required for activity. This may indicate that the overall structure of this cysteine motif is important for activity.

Interestingly, all of these singly modified toxins which we have analyzed bind red cell membranes and form oligomers in membranes as efficiently as the wild type toxin. Although these assays were not strictly quantitative, we believe a twofold difference in binding and/or oligomer formation would have been detected. The reason why some of these modified toxins exhibit dramatically reduced cytolytic activity remains unclear. Perhaps the nature of the transmembrane pores formed by these mutant toxins are defective. This is presently under investigation.

Preliminary experiments using the Glu_{434}>Asp and the double mutant Cys_{428}>Gly, Trp_{433}>Phe indicate that they may have a reduced capacity to bind cholesterol. It is thus tempting to speculate that the 11 amino acid cysteine motif is responsible (at least in part) for interaction with sterols. Interestingly the -WEWW-sequence is also found in a human sterol hydrolase, although the significance of this observation remains unclear.

Whether or not the 11 amino cysteine motif is the receptor recognition domain of the toxin remains unproven. It is interesting to note that although the Glu_{434}>Asp mutant apparently exhibits reduced ability to bind free cholesterol we could detect no reduction in its ability to bind red cells. It thus remains possible that the cysteine motif is involved in sterol interaction <u>after</u> binding to a receptor distinct from cholesterol via a toxin domain different from the cysteine motif.

```
Pneumolysin      364 - L L D H S G A Y Y A Q -  374
Listeriolysin    420 - L I D H S G G Y V A Q -  430
Perfringolysin   345 - N L D H S G A Y V A Q -  405
Streptolysin O   469 - N L S H Q G A Y V A Q -  479
                              ↑
```

Fig. 2.   The common histidine residue of the thoil-activated
           toxins. The best fit alignment of the sequence of
           four thiol-activated toxins reveals only a single,
           common histidine residue.

THE HISTIDINE MOTIF

     Treatment of pneumolysin with Diethyl pyrocarbonate
(DEPC) effectively abolished the cytolytic activity of the
toxin and its ability to bind red cells. It is therefore
tempting to speculate that such chemical modifications affect
the receptor binding domain. Since DEPC primarily modifies
histidine residues and given that each of the sequenced
toxins share extensive homology and probably a common mode of
action, we analyzed histidine residues in all of the toxins
so far sequenced. Interestingly, only a single histidine in
each of the toxins, which is outside the cysteine motif, is
aligned when the toxins sequenced are aligned for maximal
sequence similarity (see Fig. 2). This   raises the interesting hypothesis that it is this histidine residue which comprises part of the receptor binding domain of the protein. We
have changed this histidine to arginine by site-directed
mutagenesis. The mutant toxin was devoid of cytolytic activity. We are presently purifying this modified toxin in order
to carry out red cell and cholesterol binding assays.

     All of the mutants in the cysteine motif we have ana-
lyzed activate complement as efficiently as the wild-type
toxin regardless of their cytolytic activities. This implies
that the cysteine motif is not the region of pneumolysin
responsible for complement activation.

COMPLEMENT ACTIVATION

     During the production of monoclonal antibodies to pneu-
molysin we noticed that the toxin bound efficiently the secondary conjugated antibody used in the ELISA-screening method
employed for identification of hybridomas. We subsequently
showed that the Fc fragment and not the Fab fragments of IgG
bound to pneumolysin. This may account for the ability of
pneumolysin (and perhaps the other thiol-activated toxins) to
activate complement in the absence of anti-toxin antibodies.

     We have attempted to localize the Fc binding domain of
pneumolysin. A search of protein sequence data bases for
sequences similar to pneumolysin reveals little of substantial similarity. The most similar protein was human C-react-
ive protein (fig 3). This acute phase protein was first
identified by its ability to bind the C - polysaccharide of

```
Human CRP      158   GGPFSPN-VLNWRALKYEVQG-EVFTKPQLW
                     .|.. ..  ..| .|.|. || || | |. |
Pneumolysin    368   SGAYVAQYYITWDELSYDHQGKEVLT-PKAW
```

| • identity

. • conserved residue

Fig. 3.   The CRP-like domain of pneumolysin. The sequences
          of human CRP and pneumolysin were aligned using the
          FASTP program of Lipman and Pearson.

pneumococci and then by its ability to activate complement
(18). In addition, several groups have found that CRP will
protect mice from challenge with virulent pneumococci
(19,20,21). Thus it was intriguing that pneumolysin was
similar to this host protein in terms of sequence and biological properties. This may indicate that pneumolysin may have
a role in infection by competing with CRP and diverting the
products of normal CRP action from the site of infection.
This is now a testable hypothesis.

On the assumption that the common region in pneumolysin
and human CRP is that which mediates complement activation
perhaps via non- immune binding of antibody via Fc, we have
carried out site directed mutagenesis of this region in pneumolysin (Table 2). All the three replacements made affect the
ability of pneumolysin to activate complement relatively to
the native toxin. In the case of Tyr_{384}>Phe this reduced
ability to activate complement is mirrored by a reduction in
Fc binding. We tentatively propose that the region between
residues 368-397 of pneumolysin is an Fc-binding domain, and
that non-immune binding of antibody via Fc results in activation of the classical pathway of complement. This model also
predicts that the similar domain in CRP has the same properties.

IN VIVO ROLE OF PNEUMOLYSIN

A definitive role for pneumolysin in the pathogenesis of
pneumococcal disease has not yet been established, although
several lines of evidence point to an involvement. Firstly,
patients and carriers of pneumococci have anti-pneumolysin
antibodies and there is a rise in anti-pneumolysin antibodies
during pneumococcal infection (22, unpublished data). Secondly, immunization of mice with pneumolysin inactivated by
oxidation confers some protection to nasal challenge with
virulent pneumococci (23). Mice immunized in this way succumb
to the infection some 4-5 days after the non-immunized controls. The basis for this protection is not yet clear.

The above observations suggest that pneumolysin might be
a valuable addition to the existing pneumococcal capsular
polysaccharide vaccine perhaps being exploited as a protein
carrier for polysaccharide. Pneumolysin is produced by all
serotypes of the pneumococcus and alone confers limited
protection in a serotype independent manner (23). Conjugated
to the existing multivalent polysaccharide vaccine, the toxin
might also induce B cell memory to polysaccharide antigen and

Modification	Haemolytic Activity (%)	Complement Activation (%)	Fc binding (%)
Trp $_{379}$>Phe	100	55	ND
Tyr $_{384}$>Phe	100	14	25
Trp $_{397}$>Phe	100	80	ND
Native	100	100	100
Recombinant	100	100	100

Table 2. Mutagenesis of the CRP-like domain of pneumolysin-Hemolytic activity and complement activation were measured as described in Table 1. F_c binding was measured by incubating toxin with purified F_c and measuring F_c bound by ELISA.

improve the efficacy of the vaccine in the very young. Recombinant pneumolysin is the most logical source of the toxin for this purpose. Firstly, it is available in very large amounts via high level expression in *E. coli* (13) and secondly, genetically engineered toxoids are already available (see above). We have recently constructed a triple mutant with dramatically reduced cytolytic and complement activating activity. Mice were immunized with native toxin or the Cys$_{428}$>Gly mutant and challenged intra-nasally with virulent type II pneumococci. The non-immunized animals died within 3 - 4 days whilst some of the vaccinates animals survived for up to 5 - 9 days. We are currently pursuing this study using other genetically engineered toxoids whose lack of activity permits vaccination with more protein.

To explore the basis of the partial protective effect of pneumolysin, we have sought to refine the present animal model for pneumococcal pneumonia/septicaemia. Rather than using death of the animal as the experimental end point, we performed a time course following infection monitoring appearance of pneumococci in various organs following intra nasal challenge with type III pneumococci. Our preliminary results indicate that for about 40 hours post challenge pneumococci are found only in the lungs. Those animals were healthy and the lungs showed no obvious sign of consolidation. At 44 hrs post infection bacteria appear in the blood, spleen, liver and brain and following a period of rapid pneumococcal growth the animals die within about 10 hours. The events which result in translocation from the lung and explosive growth in other organs are now a topic for detailed study. It will be interesting if immunization with pneumolysin extends the interval of apparent disease-free colonization of the lung.

The above results indicate that pneumolysin might have a role to play in the pathogenesis, although by which mechanism is unclear. Clearly interference with PMNL function might be crucial to the outcome of pneumococcal infection given the central role these cells play in host defence (10). Alternatively, the interplay with CRP and complement may act to reduce their already demonstrated protective capacity.

In collaboration with Dr Rob Wilson (Brompton Hospital, London) we have demonstrated that pneumolysin is a potent inhibitor of cilia beating at least in *in vitro* organ cultures of ciliated human respiratory tract epithelia (25). Treatment of such normal epithelia results in disruption of cilia beat frequency and extrusion of cilia and membrane blabbing from the epithelia. This is not a consequence of disruption of cilial structure since in cross-section the cilia have their normal appearance. Thus pneumolysin may act to compromise the normal, non-specific defenses in the lung allowing access of pneumococci to alveoli. In this location pneumolysin may interfere with the antimicrobial activity of resident macrophages and following infiltration of PMNLs, complement and CRP act to abrogate their protective capacities.

ACKNOWLEDGMENTS

We thank Drs James Paton and Rob Wilson for their contribution to this work. Work in Leicester was supported by grants from the MRC and the Lister Institute.

REFERENCES

1. Smyth, C.J. and J.L.Duncan. 1978. In Bacterial Toxins and Cell Membranes. J. Jeljaszewice and T. Wadstrom, Eds. pp.129-183. Academic Press, London.
2. Bhakdi, S. and J. Tranum-Jensen. 1986. Microb. Pathogen. 1:5-14.
3. Alouf, J.E. 1980. Pharmac. Ther. 11:661-717.
4. Geoffrey, G., A.-M. Gilles and J.E. Alouf. 1981. Biochem. Biophys. Acta 99:781-788.
5. Mengaud, J., M.-F. Vincente. J. Chenevert, J.M. Pereira, C. Geoffroy, B. Gicquell-Sanzey, F. Baquero, J.-C. Perez-Diaz and P. Cossart. 1988. Infect. Immun. 556:766-771.
6. Duncan, J.L. and R. Schlagel. 1975. J. Cell. Biol. 67:160-173.
7. Bhakdi, S.,J. Tranum-Jensen and A. Szieysleit. 1985. Infect.Immun. 47:52-60.
8. Bhakdi, S. and J. Tranum-Jensen. 1987. Rev.Physiol.-Biochem.Pharmacol. 107:147-227.
9. Paton, J.C. and A. Ferrante. 1983. Infect.Immun. 41:1212-1216.
10. Ferrante, A., B. Rowan-Kelly and J.C. Paton. 1984. Infect.Immun. 46: 585-589.
11. Paton, J.C., B. Rowan-Kelly and A. Ferrante. 1984. Infect.Immun. 43, 1085-1087.
12. Walker, J.A., R.L.Allen, P. Falmagne, M.K.Johnson and G.J. Boulnois. 1987. Infect.Immun. 55 :1184-1189.
13. Mitchell, T.J., J.A. Walker, F.K. Saunders, P.W. Andrew and G.J. Boulnois. 1989. Biochem.Biophys.Acta. 1007, 67-72.
14. Kehoe, M.A., L. Miller, J.A. Walker and G.J. Boulnois. 1987. Infect.Immun. 55: 3228-3232.
15. Tweten, R. 1988. Infect.Immun. 56: 3235-3240.
16. Johnson, M.K. 1977. F.E.M.S. Microbiol. Lett. 2:243-245.
17. Saunders, F.K., T.J. Mitchell, J.A. Walker, P.W. Andrew and G.J. Boulnois. 1989. Infect.Immun. In Press.

18. Edwards, K.M., C. Mold, R. F. List, H. Gewurz. 1980. Clin.Res. 28:735A.
19. Horowitz, J., J.E. Volonakis and D.E. Briles. 1987. J.Immunol. 138:2598-2603.
20. Mold, C., S. Nakayama, T.J. Holzer, H. Gewurz and T.W. Du Cols. 1981. J.Exp.Med. 154: 1705-1708.
21. Yother, J., J.E. Volanakis and D.E. Briles. 1982. J.Immunol. 128, 2374-2376.
22. Kalin, M., K. Kanderski, M. Granstrom and R. Mollby. 1987. J.Clin.Microbiol. 25:226-229.
23. Paton, J.C., R.A. Lock and D.J. Hansman. 1983. Infect.Immun. 40:548-552.
24. Steinfort, C., R. Wilson, T. Mitchell, C. Feldman, A. Rutman, H. Todd, D. Sykes, J.Walker, K. Saunders, P.W. Andrew, G.J. Boulnois and P.J. Cole. 1989. Infect.Immun. In Press.

STUDIES ON THE GENETIC BASIS OF *HAEMOPHILUS*

INFLUENZAE PATHOGENICITY

E. Richard Moxon and Jeffrey N. Weiser

Molecular Infectious Diseases Group, Oxford University
Department of Paediatrics, Institute of Molecular
Medicine, John Radcliffe Hospital, Headington, Oxford
OX3 9DU, United Kingdom

INTRODUCTION

The pathogenesis of bacterial meningitis comprises a sequence of several distinct events; acquisition of the causative bacterium from another individual, colonization, invasion of mucosal epithelium resulting in bacteremia, haematogenous dissemination, invasion of the blood-brain barrier and, finally, the occurrence of inflammatory changes within the subarachnoid space consequent upon bacterial multiplication within the central nervous system (Figure 1).

Over a number of years, my group has investigated the molecular basis of *H. influenzae* pathogenesis by combining genetic techniques, which define microbial virulence determinants, and a convenient, biologically relevant experimental model of meningitis in rats. Our studies have indicated the central importance of bacteremia as a determinant of meningeal invasion and that *H. influenzae* is capable of efficient intravascular replication (1). Since the occurrence of meningitis depends in part upon the magnitude and duration of bacteremia, the microbial factors which are required for efficient multiplication and survival in the blood are of particular importance to understanding the potential of *H. influenzae* to cause CNS invasion (2).

Several potential virulence determinants of *H. influenzae* have been identified (Table) and their relative roles can be analyzed *in vitro* and *in vivo* using molecular genetic techniques to clone the relevant genes and to construct strains which express, or do not express, specific determinants. It has been shown that the capsular polysaccharides and the lipopolysaccharides of *H. influenzae* are important in virulence expression (3,4).

More than 50 years ago, Margaret Pittman recognized that virtually all cases of invasive H. influenzae disease were caused by capsulate strains (5); of the 6 antigenically distinct capsular serotypes, designated a-f, strains express-

Microbial Surface Components and Toxins in Relation to Pathogenesis
Edited by E.Z. Ron and S. Rottem, Plenum Press, New York, 1991

171

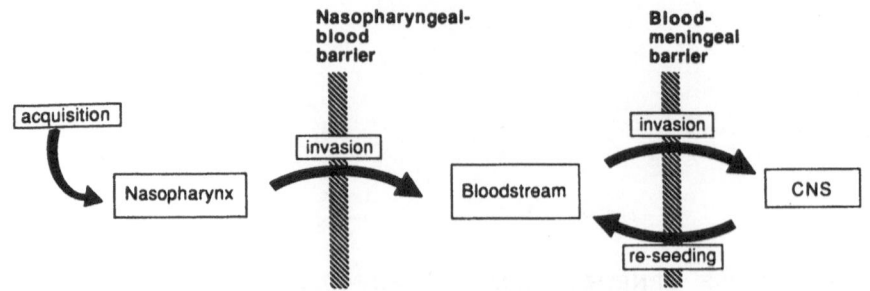

Figure 1. Pathogenetic sequence for bacterial meningitis caused by *Haemophilus influenzae* type b.

ing the type b poly-ribosyl-ribitol phosphate polysaccharide capsule count for greater that 90% of invasive infections, including meningitis (6). The cloning of the genes for type b capsule has made it possible to assess the contribution of the different capsular polysaccharides to the virulence of *H. influenzae* in the rat model of meningitis. A series of transformants, each expressing a different capsular polysaccharide, has been constructed. Physical mapping of the transformants has shown that each differs specifically in that region of the chromosome where the DNA conferring type specificity is located (7). The transformants share apparently identical phenotypic characteristics including outer membrane proteins and lipopolysaccharides; they also share an identical genotype based on an analysis of metabolic enzymes using the technique of multi-locus enzyme electrophoresis. Following intranasal challenge of infant rats, only the serotype a and b transformants consistently caused bacteremia and the type b transformant was significantly more virulent (8). Interestingly, both type a and type b polysaccharides include ribitol phosphate, which is not a component of the four other capsules. The comparative virulence studies provide strong evidence that type b capsule genes are critical for maximal virulence. In addition to its role in preventing clearance of *H. influenzae* from the blood, recent studies indicate that type b capsule influences meningeal invasion and alters blood-brain permeability since these correlate directly with the number of surviving organisms in the subarachnoid space (9). However, when compared to the other capsular polysaccharides, the relative contribution of type b capsule to nasopharyngeal colonization and cellular invasion has not been determined.

Although the contribution of virulence of the lipopolysaccharides of many gram-negative bacteria has been documented, a definite role in *H. influenzae* pathogenicity was not clearly demonstrated until more recently. The LPS of *H. influenzae* is rough, resembling that of *Neisseria* and *Bordetella* and is referred to by many investigators as lipo-oligosaccharide (10,11). Unlike the enterobacteriacae, whose LPS molecules possess repetitive oligosaccharide units, those of *H. influenzae* lack these repetitive structures and possess only the core region comprising short chains of covalently linked neutral sugars. At present, there are no details of the structure of the *H. influenzae* core region. However, the composition of the core oligosaccharides of H. influenzae differs from those of *Salmonella*, *Shigella* or *E.coli*. When

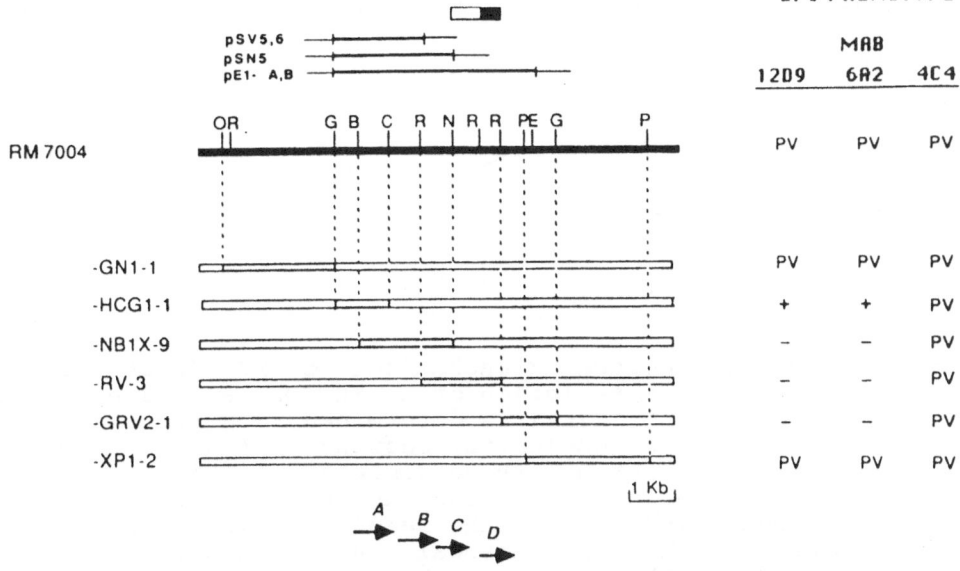

Figure 2. The phenotypic effect of deletion/insertion mutagenesis in the *lic*-1 region. The physical map of
the chromosomal region encompassing *lic*-1 in strain
RM.7004 is shown as a solid bar. Restriction endo-
nucleases: B (*BamH*l), C(*Cla* 1), E(*Eco* R1) G(*Bgl*
II), N(*Nru* 1), O (*Nco* 1), P(*Pst* 1) and (*Eco* RV).
Several deletion/insertion mutants (deleted segment
stippled) are indicated below the map of RM.7004.
The phenotype of RM.7004 and the mutants (reactivity with mabs 12D9, 6A2 and 4C4) are indicated to
the right of the maps. PV indicates that the mutant
reacted with the mab and showed phase-variation;
(+) constitutive reactivity and (−) constitutive
nonreactivity. Arrows at the bottom of the figure
indicate the position four genes (l A-D) comprising
the *lic*-1 operon which is transcribed from left to
right. Above the map of RM.7004, the plasmid in-
serts used in minicell protein expression system
are shown. The locations of sequences previously
shown to possess transforming activity for the 12D9
epitope (solid box) and 6A2 epitope (open box) are
indicated above the map of RM.7004

H. influenzae LPS was extracted and analyzed by SDS-PAGE and
silver staining, substantial heterogeneity of the molecules
obtained from different strains was found (11). The availability of monoclonal antibodies to the oligosaccharide (12)
has defined several epitopes and these have provided a basis
for sub-typing. Furthermore, using colony immuno-blotting,
individual strains show antigenic variation in which there is
spontaneous and reversible loss of oligosaccharide epitopes
at high frequency (approximately 10^{-2} per generation), a
phenomenon which is typical of phase-variation. Differences
in virulence were found when rats were challenged with
organisms which were predominantly in one phase or the

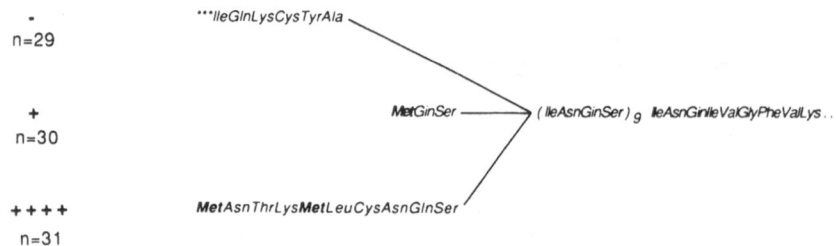

AATATAAAA[ATG]AATACAAAA[ATG]CT[ATG] .... (CAAT)n .... CAAATTGTAGGATTTGTTAAA

PHENOTYPE

- n=29

···IleGlnLysCysTyrAla

+ n=30

MetGlnSer ⟶ (IleAsnGlnSer)₉ IleAsnGlnIleValGlyPheValLys...

++++ n=31

MetAsnThrLysMetLeuCysAsnGlnSer

Figure 3. Proposed molecular mechanisms of LPS phase-variation. Phase-variation of *H. influenzae* LPS epitopes expressed by *lic*-1 is determined by a translational switch. The nucleotide sequence of the 5' end of *lic* A is listed with possible ATG initiation codons boxed. Variation in the number of multiple CAAT repeats (n) shifts the reading frame of *lic* A altering the translational phase up-stream of the repeats. The three possible amino-terminal translational products (n = 29-31) are shown below the nucleotide sequence. Depending upon the number of CAAT units there are either one, two or no initiation codons (Met indicated in bold lettering) in frame causing variable translation of *lic* A. It is proposed that the three levels of expression observed for *lic*-1 determined epitopes (+, ++++, -) corresponds to the three translation phases at the 5' end of *lic* A.

other (12). In order to clarify the role of oligosaccharide epitope expression in *H. influenzae* virulence, Weiser made a genomic library from a virulent type b strain and screened it for clones which could transform a recipient *H. influenzae* strain which was constitutively lacking in the expression of specific oligosaccharide epitopes expressed by the donor strain. Transformants expressed novel epitopes which were defined by reactivity with monoclonal antibodies. Further studies have defined a chromosomal locus (*lic*1) from which subcloned DNA was used to generate a series of isogenic variants which expressed different oligosaccharide epitopes. Transformants which acquired novel epitopes displayed phase--variation of these oligosaccharides; also the altered phenotype correlated with changes in the composition of the core sugars. By introducing site-specific (frame-shift) mutations into sub-cloned DNA from *lic*1 and introducing this DNA into the present strain by transformation, loci corresponding to two specific epitopes were identified (13). The DNA sequence of *lic*1 has shown that it comprises a single transcriptional unit of 4 genes. The first of these genes (*lic*A) is involved in phase-variation since deletions or frame-shift mutations within it result in constitutive expression of the two oligosaccaride epitopes mediated by genes (*lic*C and *lic*D) within *lic*1 which are downstream from *lic*A (Figure 2). At the 5' end of *lic*A, within the open reading frame, are many tandem repeats of the tetramer CAAT. There are 3 possible ATG initi-

174

ation codons 5' to the repeat. When *lic*A was sequenced initially in the M13 bacteriophage vector, the insert containing the tandemly repeated copies of CAAT had 29 repeats and when sequenced subsequently in a pUC19 vector, only 26 repeats were found (14). Translation of the sequence with either 26 or 29 repeats, a difference of 12 base pairs, results in all possible ATG initiation codons being out-of-frame. In contrast, 30 repeats of CAAT would place an ATG (found immediately adjacent and upstream of the repeats) in frame (Figure 3). Likewise, if there are 31 repeats of CAAT, either of the two other ATG start codons would be in-frame for translation of *lic*A. These findings suggest that the tandem repeats of CAAT at the 5' end of *lic*A provide a mechanism for the phase-variation. Changes in the number of copies of CAAT (e.g. loss or gain of one copy - perhaps through slipped- strand mispairing) would shift the three initiation codons in and out of frame with the remainder of the *lic*A open reading frame. Since *lic*A is required for phase-variation, the repeats would create a translational switch for this critical protein. To test whether changes in the number of CAAT repeats correlated with altered phenotypic expression of the oligosaccharides, the polymerase chain reaction was used to amplify segments of DNA containing the CAAT repeats. The most prominent DNA species obtained from negative variants (i.e. those failing to react with monoclonal antibodies specific for an oligosaccharide epitope) contain 29 repeats whereas a strongly positive reactive variant yielded DNA molecules of the size corresponding to 31 repeats (14). A novel feature of the phase-variation is that when the switch is 'on', there are 2 levels of expression of the relevant oligosaccharide epitopes. Altogether, three levels of expression (⁻, ⁺, and ⁺⁺⁺⁺) for *lic*1 controlled epitopes occur corresponding to the situation depicted in Figure 3. Interestingly a similar translational switch has been reported for the control of the *opa* genes in *Neisseria* (15); these genes are also constitutively transcribed, but not all of the transcripts are translated into functional proteins. This regulation was found to be the result of a repetitive DNA motif (7-28 copies of CTCTT) within the DNA sequence coding for the hydrophobic portion of the *opa* signal sequence (15).

The *lic*1 translational switch of *H. influenzae* could account for the phase-variation of only 2 of the several oligosaccharide epitopes which bind different monoclonal antibodies. In addition to *lic*1, strain RM.7004 has two other chromosomal loci (*lic*2 and *lic*3) containing multiple tandem repeats of CAAT. A survey of other *H. influenzae* strains (encapsulated and non-encapsulated), had indicated that *H. influenzae* has from 2-5 chromosomal loci containing repeats of CAAT. Recently, two additional loci (lic2 and *lic*3) from RM.7004 have been cloned. Similar to *lic*1, mutations in *lic*2 result in altered expression of oligosaccharides which are defined by monoclonal antibodies, but the role of *lic*3 has yet to be defined. Using pulse field gel electrophoresis, the *lic*1-3 loci have been mapped and lie within a single 215 kb fragment generated by digestion of the type b chromosome using the endonuclease *Eag* I (personal communication: Dr. Peter Butler).

The data presented currently indicate that the expression of the core oligosaccarides of *H. influenzae* is regulat-

ed by switch mechanisms so as to result in substantial diversity of immuno-dominant epitopes. It has been shown that
on-off switching of one individual epitope occurs independent
of, or concomitant with, that of other epitopes. This implies
that *H. influenzae* core sugars have the potential to generate
an extensive repertoire of different structures. This antigenic variation, easily demonstrated *in vitro* (12,13), provides a mechanism which could enhance both the commensal and
pathogenic behaviour of *H. influenzae* in man. The induction
or selection of variants could facilitate, for example, the
organisms' ability to evade host immune responses or increase
the number of organisms expressing a structure which favours
attachment to host cells. Phase-variation does apparently
occur *in vivo*. Immuno-peroxidase staining of organisms in the
cerebrospinal fluid obtained from children with bacterial
meningitis showed that most of the organisms were in the same
phase (either 'on' or 'off') as evidenced by their reactivity
with monoclonal antibodies; however, occasional organisms
were in the opposite phase (14). This finding was consistent
with the occurrence of phase-variation within a bacterial
population which included a majority of organisms in which
induction or selection of a particular epitope has occurred.

Evidence that phase-variation may be biologically relevant to the pathogenicity of type b strains has been obtained
through the experimental infection of infant rats. By introducing mutations into *lic*1 and *lic*2, the expression of oligosaccharide epitopes controlled by these loci was abolished.
This strain was compared to the wild-type parent. Following
intranasal challenge, the rats inoculated with the wild-type
organisms developed bacteremia significantly more frequently
as compared to the mutant. Thus, expression of particular
epitopes and/or the capacity to undergo phase-variation is
evidently potentially important in the pathogenesis of *H.
influenzae* b meningitis.

The antigenic variation of oligosaccharide epitopes of
H. influenzae may explain much of the phenotypic heterogeneity which has been observed by previous investigators (10--
12). Also, both colony blotting and the use of a simple
immunoperoxidase stain of single bacterial cells indicates
that there are broad similarities between the phenomenon of
oligosaccharide phase-variation in *H. influenzae* and that
described independently for *Neisseria* LPS (16). Both of these
species of bacteria are commensals, as well as important
pathogens of man; it is intriguing that both have evolved
mechanisms which generate extensive antigenic switching and
diversity of their core sugars. Recently, it has been shown
that the LPS of *Haemophilus* and *Neisseria* share many common
oligosaccharide epitopes not present in the LPS of other
gram-negative bacteria. One of these structures is `-D-galac-
topyranosyl-(1-4)-a-D-galacto-pyranose (17), a structure
found on human epithelial cells which is a receptor for the
Escherichia coli Pap adhesin (18) and the enterotoxin of
Shigella (19). Although *Haemophilus and Neisseria* have
evolved similar mechanisms which effect rapid antigenic
switching of core oligosaccharides, there was no hybridisat-
tion of LPS genes of *H. influenzae* to the genome of *N. gonorrhoea* or *meningitidis*. These findings suggest that these
genetically distinct genera may have evolved common LPS core
structures through convergent evolution; future studies on

Haemophilus and on the genetics of LPS expression in *Neisseria* should prove interesting in defining the similarities and differences in these molecular mechanisms of antigenic variation.

ACKNOWLEDGEMENTS

The authors wish to acknowledge the contributions of Drs. P. Butler, E.J. Hansen, J.S. Kroll, A. Lindberg, D. Maskell, M. Virji and A. Williams to these investigations, and Mrs. Seila Hayes for her expertise in the preparation of the manuscript.

The work was supported by funding from the Medical Research Council and The Meningitis Trust.

REFERENCES

1. Rubin, L.G., A. Zwahlen and E.R. Moxon. 1985. J. Infect. Dis. 152:307-314.
2. Moxon, E.R. and J.S. Kroll, J.S. 1989. In:Bacterial Capsules & Adhesins: Facts & Principles. Edited by K. & B. Jann. Springer Verlag pp. 65-85.
3. Ely, S., J. Tippett, J.S. Kroll, and E.R. Moxon, E.R. 1986. J. Bacteriol. 167:44-48.
4. Zwahlen, A., L.C. Rubin and E.R. Moxon. 1986. Microb. Pathog. 1:465-473.
5. Pittman, M. 1931. J. Exp. Med. 53:471-793.
6. Turk, D.C. 1982. In: S.W. Sell and P. Wright, eds. *Haemophilus* New York: Elsevier Science Publishing Co.:Inc.:p3-9.
7. Kroll, J.S., S. Zamze, B. Loynds and E.R. Moxon. 1989. J. Bacteriol. 171:3343-3347.
8. Zwahlen, A., J.S. Kroll, L.G. Rubin and E.R. Moxon. 1989. Microb. Pathog. 7:225-235.
9. Lesse, A., W.M. Scheld, A. Zwahlen and E.R. Moxon. 1988.. J. Clin. Invest. 82:102-109.
10. Inzana, T.J. 1983. J. Infect. Dis. 148:492-499.
11. Zamze, S.E. and E.R. Moxon. 1987. J. Gen. Micro. 133:1443-1451.
12. Kimura, A., and E.J. Hansen. 1986. Infect. Immun. 51:69-79.
13. Weiser, J.N., A.A. Lindberg, E.J. Manning, E.J. Hansen and E.R. Moxon. 1989. Infect. Immun. 57:3045-3052.
14. Weiser, J.N., J.M. Love and E.R. Moxon. 1989. Cell 59:657-665.
15. Stern, A., M. Brown, P. Nickel and T.F. Meyer. 1986. Cell 47:61-71.
16. Schneider,H., T.L. Hale, W.D. Zollinger, R.C. Seid, Jr., C.A. Hammack and J. McL. Griffiss. 1984. Infect. Immun. 45:544-549.
17. Virji, M., J.N. Weiser, A.A. Lindberg and E.R. Moxon. 1989. Microb. Path. (in press).
18. Lindberg, F.P., B. Lund and S. Normark. 1984. EMBO J. 3:1167-1173.
19. Lindberg, A.A., J.E. Schultz, M. Westling, J.E. Brown, S.W. Rothman, K.-A. Karlsson and N. Stromberg. 1986. In: D.L.Lard:ed.: Protein-Carbohydrate Interactions in Biological Systems. Academic Press:Inc. p361-367.

IV. TOXINS AND SYSTEMATIC EFFECTS

ASSEMBLY AND SECRETION OF OLIGOMERIC TOXINS

Timothy R. Hirst[1], Maria Sandkvist[2],
Robert Aitken[1], and Michael Bagdasarian[2]

[1]Department of Genetics, University of Leicester
Leicester LE1 7RH, Great Britain and [2]Michigan
Biotechnology Institute, Lansing, MI 48909, USA

INTRODUCTION

Pathogenic microorganisms produce a myriad of different virulence factors that range in complexity from polymers of simple disaccharides to multimeric protein toxins and adhesins. This structural and functional diversity belies a common biosynthetic requirement shared by all virulence factors: that they must cross the membrane of the bacterium in which they are produced before they can gain access to the host and express their pathogenic properties. The processes of membrane translocation result in either the complete secretion of a pathogenic molecule into the external milieu, eg. extracellular toxins, siderophores etc., or to their assembly onto the bacterial surface, as in the case of adhesins, pili, outer membrane porins, and carbohydrate capsules. In this paper we discuss the mechanisms of protein export and secretion in bacterial cells, and in particular the secretion of complex oligomeric toxins that are responsible for causing cholera and related diarrhoeal diseases.

PROTEIN EXPORT

The phenomenon of protein export occurs in all cells and has been studied in a range of prokaryotic and eukaryotic systems (for reviews, see refs 1-4). Production and secretion of oligomeric toxins have been found to occur only in Gram-negative bacteria. Gram-negative bacteria have complex envelope structures, consisting of an inner (cytoplasmic) membrane and an outer membrane separated by an aqueous periplasmic compartment (3). The mechanism of protein export through the cytoplasmic membrane is remarkably similar to the processes of protein translocation across the ER of eukaryotic cells and to the import of proteins into mitochondria and chloroplasts (5). Almost all of the proteins that cross the cytoplasmic membranes of bacteria do so by this common mechanism: the main features of which include: synthesis of the exported protein as a longer precursor with an amino-terminal signal sequence (4,6); the coupling of the precursor via specific factors to an export apparatus in the

Microbial Surface Components and Toxins in Relation to Pathogenesis
Edited by E.Z. Ron and S. Rottem, Plenum Press, New York, 1991

181

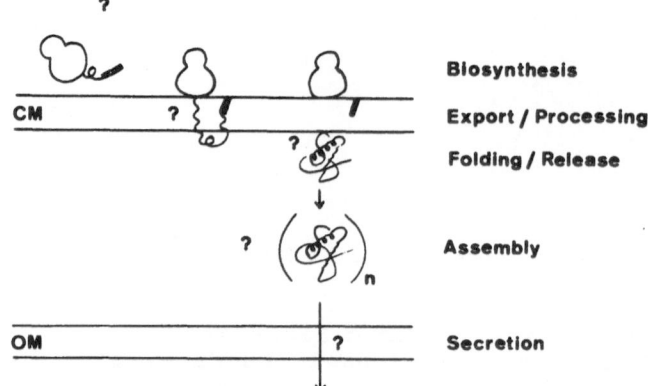

	Biosynthesis
CM	Export / Processing
	Folding / Release
	Assembly
OM	Secretion

Fig. 1.   Pathway of protein secretion from Gram-negative
bacteria. Almost all proteins that are exported
from Gram-negative bacteria are produced with
amino-terminal signal sequences. Such precursors
interact with soluble cytoplasmic factors and an
export apparatus in the cytoplasmic membrane.
After translocating across the membrane, the signal
sequence is cleaved off and the protein folds in a
soluble conformation and is released from the membrane. For some proteins, eg oligomeric toxins,
polypeptide subunits assemble within the periplasm
and the molecule is secreted across the outer membrane. The aspects of protein export and secretion
that remain to be understood at a molecular level
are indicated by question marks.

membrane (7-11); translocation across the membrane by an
energy-driven mechanism that requires ATP and an electro-
chemical potential (12); removal of the signal sequence by
specific proteases, and folding and release of the protein on
the other side of the inner membrane (13-14) (Fig. 1). The
latter step of protein folding and release from the membrane
is particularly obscure, and will be discussed in relation to
the secretion of diarrhoeagenic toxins.

Subsequent events in protein secretion, that follow
after the release step from the cytoplasmic membrane, are
dependent on the particular protein and the microorganism.
In Gram-positive bacteria, for example, the absence of an
outer membrane means that exported proteins may diffuse
freely into the surrounding milieu. In Gram-negative bacteria
however, the outer membrane represents an important additional barrier through which secreted products must pass (Fig.
1). Porins and other proteins which are located in the outer
membrane itself, appear to be able to fold into structures
that allow them to partition into the membrane, whereas
surface appendages and extracellular toxins, have evolved a
variety of outer membrane translocation mechanisms (3).

The most straight-forward is exemplified by the mechanism for secretion of IgA protease from *Neisseria gonorrhoeae,* in which the mature enzyme, after insertion into the
outer membrane is autoproteolytically cleaved and released

Table 1   Alterations at the carbolxl-teminus of EtxB

B subunit	C-terminal amino acids	Sequence
EtxB	W.T.	...IleSerMetGluAsn^{+103}
EtxB191.5	-4 +3	...IleGlyLeuAsn
EtxB214	-1	...IleSerMetGlu
EtxB215	-2	...IleSerMet
EtxB216	-3	...IleSer
EtxB124	-1 +7	...IleSerMetGluLysLeuPheGlnProAspThrAsp
EtxB138	-1 +7	...IleSerMetGluLysLeuAlaProGlnLysArgTrp

into the surrounding milieu (15). When IgA protease was cloned into *Escherichia coli* it was found to be efficiently secreted from this organism as well, thus indicating that no accessory (sec) factors from *N. gonorrhoeae* are needed to effect protease secretion (15). However, for other extracellular proteins membrane translocation appears to be much more complex. For example aerolysin from *Aeromonas hydrophila* (16), cholera toxin from *Vibrio cholerae* (17-18), and exotoxin A from *Pseudomonas aeruginosa* (19) are not secreted into the external milieu when expressed in *E. coli*, and therefore require the activity of "Sec" factors from the membranes of their parent organisms in order to pass through the outer membrane.

OLIGOMERIC ENTEROTOXINS

Our studies on oligomeric enterotoxins have provided a convenient model for investigating each stage of the secretion process of Gram-negative bacteria. These proteins are multimeric complexes consisting of six polypeptides assembled together by non-covalent interactions (20). Each toxin molecule is composed of a single A subunit (Mw:28,000), which catalyses ADP-ribosylation of adenylate cyclase and five identical B subunits (Mw:12,000) which bind the toxin to ganglioside receptors on the surfaces of host tissues. Most of our studies have been performed using the heat-labile enterotoxin (LT) produced by toxinogenic strains of *E. coli*, which shows considerable structural, functional and immunological homology to cholera toxin, the prototype of this family of enterotoxins (21). The A and B subunits are encoded by a single polycistronic mRNA which specifies the synthesis of precursor A and B subunits with 18 and 21 amino acid signal sequences respectively (22,23). Entry and translocation of the precursors across the cytoplasmic membrane of either *E. coli* or *V. cholerae* results in the release of the mature A and B subunits into the periplasm (17,18).

Table 2   Stability of C-terminally deleted EtxB

B subunit	C-terminal amino acids	Subunit concentration (ng/ml)	
		CC118	KS476 (degP)
EtxB	W.T.	4020	4070
EtxB215	-2	525	2760
EtxB191.5	-4 +3	156	490

SUBUNIT FOLDING & RELEASE FROM THE CYTOPLASMIC MEMBRANE

Previous studies by Beckwith (24), Botstein (14) and Knowles (13) on beta-lactamase and maltose binding protein, have established that there are conformational changes in exported proteins on the cytoplasmic membrane which confer upon them a soluble, protease-resistant structure.

We have been able to alter the kinetics of this conformational change in the B subunit of heat-labile enterotoxin by introducing subtle alterations at the carboxyl-terminus of the molecule (Table 1). This was achieved by Bal31 digestion and site-directed mutagenesis of the B subunit gene.

Codon alterations causing the deletion of two amino acid residues from the carboxyl terminus (see EtxB215), resulted in an approx. 8-fold reduction in the amount of detected B subunits in E. coli K-12, strain CC118 (Table 2). Deletion of four residues and their replacement with three different amino acids (EtxB191.5) had an even greater effect on B subunit stability (Table 2), and the removal >3 amino acids caused the complete loss of detectable B subunit. That these effects are due to an increased susceptibility of the mutant B subunits to proteolytic degradation was established by analyzing their expression in E. coli strain KS476 which lacks a periplasmic protease (Deg P) (25). The amounts of EtxB215 and EtxB191.5 detected in KS476 were 4-5 fold higher than in CC118, whilst the amount of native EtxB was approximately the same in the two strains (Table 2).

E. coli KS476 expressing the various deleted B subunits was radioactively pulse-labelled with ^{35}S-methionine for a period of 10 minutes and then the periplasmic and spheroplast cell fractions were isolated and analyzed by SDS-PAGE (Fig. 2). A significant proportion of the mutant B subunits were associated with the spheroplast fraction: indeed for EtxB216 which lacks the last three amino acids of the native B subunit, almost all of the radiolabelled mutant protein was in the spheroplast fraction (Fig. 2, compare lanes 4 and 8).

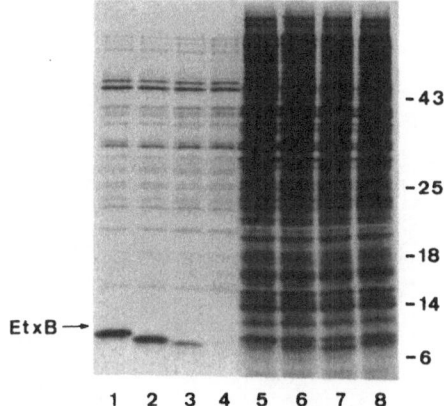

EtxB →

1 2 3 4 5 6 7 8

-43

-25

-18

-14

-6

Fig. 2. Release of C-terminally deleted B subunits from the cytoplasmic membrane in a protease-deficient (Deg P) strain of *E. coli*. Cells expressing various EtxB proteins were radiolabelled and then fractionated into periplasm (lanes 1-4) and spheroplasts (lanes 5-8) fractions, which were subjected to SDS-PAGE and autoradiography. *E. coli* KS 476 expressing native B subunit (lanes 1 and 5), EtxB215 (lanes 2 and 6), EtxB191.5 (lanes 3 and 7) and EtxB216 (lanes 4 and 8). The position of native EtxB and molecular weight markers (in kDa) are indicated.

When proteinase K was added to these fractions the periplasmically located mutant B subunits were resistant to proteolytic degradation whereas the spheroplast associated B subunits were susceptible. We therefore conclude, that subtle deletions at the carboxyl-terminus of the EtxB alter the kinetics of the B subunit folding pathway and thereby change the efficiency of subunit release from the cytoplasmic membrane.

SUBUNIT ASSEMBLY

Oligomeric toxins and polymeric surface adhesins not only have to fold up into secondary structures and tertiary domains but they must also form complementary interfaces for subunit-subunit interactions. In (hetro) oligomeric enterotoxins, this involves B subunit-B subunit association and A subunit-B subunit association.

We previously hypothesized that the assembly of B subunits into pentamers preceded their association with an A subunit (26). This was based upon the observation that B subunits assembled into pentamers in the absence of any A subunit expression. Indeed, it has even proven possible to achieve B subunit assembly *in vitro* in the complete absence of any A subunits, by dissociating highly purified B penta mers into monomers under acidic conditions and then monitoring, after neutralization, their reassembly into pentamers (Fig, 3). We have used two experimental approaches to study toxin assembly *in vitro*; (1) an SDS-PAGE analysis of the rea-

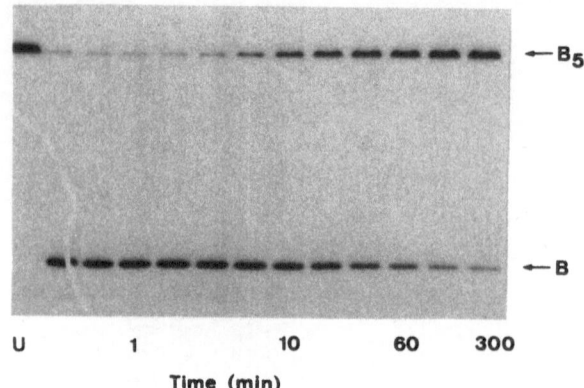

Fig. 3.   Assembly of LTB subunits *in vitro*. Purified EtxB
(0.2 mg/ml) was disassociated by addition of 100 mM
HCl. After 10 min the solution was neutralized by
addition of an equimolar concentration of NaOH, and
samples were taken into SDS-PAGE sample buffer, and
then analyzed by SDS-PAGE.   Undissociated native
EtxB pentamers (lane U) as well as the times at
which samples were mixed with sample buffer are
shown.

ssembly of acid-dissociated B monomers (27) (Fig. 3), and (2)
an immunological assay based on the use of monoclonal antibodies which react only with assembled B subunits (Fig. 4).

    Figure 3 shows that native, undissociated B subunits
migrate on an SDS-polyacrylamide gel as pentamers (lane U)
and that following acid-dissociation, B monomers reassemble
into oligomers of the same electrophoretic mobility.

    The use of the immunological assay permitted the associ-
ation of A and B subunits to be monitored by using monoclonal
antibodies against the A subunit.   It was found that the
association of A and B subunits would only occur when the B
subunits were in the process of assembling (Fig. 4). This
suggested that the A subunit normally interacts with an
intermediate in B subunit assembly (27).

    The consequence of the A subunit associating with a B
subunit intermediate is that it alters the assembly pathway.
This was found to be of particular importance when the kinetics of enterotoxin assembly were investigated *in vivo* (27).
E. coli strains expressing both A and B subunits formed
stable B pentamers approx. 3 times as rapidly as   strains
producing only the B subunits. Thus, whilst the A subunit is
not obligatory for B subunit assembly we conclude that the A
subunit is able to accelerate the rate of B subunit penta-
merization, probably by stabilizing an assembly intermediate.
The A subunit therefore plays a coordinating role in the
pathway of enterotoxin assembly (27).

MAPPING THE DOMAINS OF SUBUNIT INTERACTION

    Using a similar rationale to the one above, that subtle

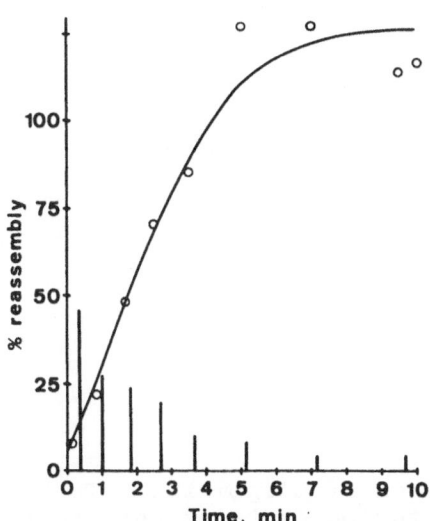

Fig. 4.   Kinetics of *in vitro* pentamer formation and A-B
interaction.   Cholera toxin B subunits (0.26 mg/ml)
were acidified with HCl and neutralized with NaOH.
At the times indicated the samples were diluted
with PBS to 0.001 mg/ml (o) or were mixed with an
equal volume of A subunit (0.13 mg/ml) and incubated for a further 10 min before dilution with PBS
(vertical bars).   The percentage of reassembled B
subunits detected with a pentamer specific monoclonal antibody (o) and the percentage of A subunits associated with B subunits using an A subunit
specific monoclonal antibody was determined using a
GM1-based enzyme-linked immunosorbant assay.   Only
A subunits that are stably associated with B subunits are detected in this assay.   (Figure taken
from ref. 27).

alterations in toxin structure may effect assembly, a variety
of mutant B subunits were engineered with short carboxy-terminal extensions (28).   Two of them, EtxB124 and EtxB138 (see
Table 1), both of which have seven extra amino acid residues,
were found to assemble into pentamers, bind to GM1-ganglio-
side receptors, and be completely secreted from *V. cholerae*
(28).   However, they were unable to assemble with A subunits
to form a hexameric holotoxin.   This implies that the carboxyl-terminal domain of the B subunit mediates A-B subunit
interaction.

TOXIN SECRETION ACROSS THE OUTER MEMBRANE OF *V. CHOLERAE*

The mechanism of toxin secretion across the outer membrane of *V. cholerae* is fundamentally different from the
process of protein export across bacterial cytoplasmic membranes.   Several observations support this conclusion: first,
it is independent of any requirement for a "classic" signal
sequence; second, the translocation step involves secretion
of a fully assembled quaternary complex (29), rather than an
unfolded polypeptide chain; third, there is no evidence for
an energy requirement; and forth, novel "Sec" factors from *V.*

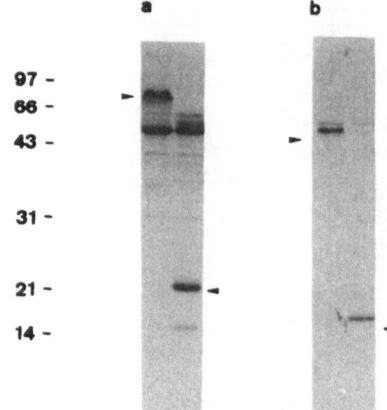

Fig. 5.   Generation of oligomeric LTB-ST vaccine antigens.
LTB-ST fusion proteins were partially purified from
the periplasm of *E. coli* G6 pTRH13 (panel a) and G6
pTRH14 (panel b). The samples were either heated to
100°C for 5 min or kept at room temperature prior
to analysis by SDS-PAGE (right and left of each
panel respectively). The LTB-ST fusion proteins are
indicated by arrows, and the position of molecular
weight marker proteins (in kDa) are shown to the
left of the figure.

cholerae are obligatory since expression of cholera and
related enterotoxins in *E. coli*, results their export only as
far as the periplasmic space (17). How *V. cholerae* accom-
plish the process of toxin secretion remains to be resolved.
This will require both an understanding of the Sec apparatus
and its role in the translocation event, and an assessment of
the toxin domains involved in this process.

We   have   observed that strains of *V. cholerae* which had
been engineered to produce only the B subunit of LT can
efficiently secrete B pentamers into the extracellular milieu, whereas strains producing only the A subunit of LT
accumulate A subunits in the periplasmic space (26). This
implies that the structural information for toxin secretion
across the outer membrane is confined to the B subunit.

Carboxyl-terminal extensions on the B subunit EtxB124
and EtxB138, did not interfere with their secretion across
the outer membrane of *V. cholerae* (28).

ENGINEERING OLIGOMERIC ENTEROTOXOID VACCINES

We are currently exploring ways of using the properties
of the B subunit enterotoxoid to engineer oligomeric multivalent vaccines.   Our observation, that carboxyl-terminal
extensions do not destabilize the B subunit meant that specific antigens or epitopes could be engineered onto the
molecule without perturbing its export, folding, assembly and
secretion properties.

The gene for heat-stable enterotoxin (ST) was spliced onto the 3'-end of the B subunit gene. Two different portions of the ST gene were used, thereby generating a longer fusion containing most of the ST gene (pTRH13) and a shorter fusion containing the 3'-toxin region alone (pTRH14).

Expression of the fusion proteins in *E. coli* resulted in their export to the periplasm and assembly into discrete oligomers (Fig. 5). An analysis of these proteins has revealed that they retain both B subunit and ST structural features, including binding to GM1-ganglioside, recognition by anti-B subunit and anti-ST antibodies, and assembly into heat-labile SDS-stable oligomers (see Fig. 5).

LTB-ST fusion proteins of this kind have considerable potential as vaccine antigens for the prevention of diarrhoeal disease, since the small size of native ST precludes its ability to induce protective immunity. The oligomeric LTB-ST proteins should circumvent this problem and allow ST immunity to be induced.

The properties of these hybrid enterotoxins indicate that the structural features of LTB involved in interaction with the export apparatus, release from the cytoplasmic membrane and assembly into pentamers are retained in each construct. It will be of interest to see if the length or conformation of the additional ST domains impair the interaction of the fusion proteins with the secretion apparatus in *V. cholerae*. Successful secretion would open up the possibility of using attenuated *Vibrio* species to deliver these antigens to the mucosal surface of the intestine.

ACKNOWLEDGEMENTS

We wish to acknowledge Simon Hardy and Jan Holmgren for their discussion and contribution of immunological reagents. TRH is a Wellcome Trust Senior Research Fellow. This work was supported by grants from The Wellcome Trust of Great Britain, The Swedish Board for Technical Development and The World Health Organization.

REFERENCES

1. Wickner W.T. and H.F. Lodish. 1985. Science 230:400-407.
2. Zimmerman R. and D.I. Meyer. 1986. Trends Biochem Science, 11:512-515.
3. Hirst T.R. and R.A. Welch. 1988. Trends Biochem Sci. 13:265-269.
4. Randall L.L. and S.J.S. Hardy. Science 243:1156-1159.
5. Randall L.L., S.J.S. Hardy and J.R. Thom. 1987. Ann Rev Microbiol. 41:507-541.
6. Ferenci T. and T.J. Silhavy. 1987. J. Bacteriol. 169:5339-5342.
7. Gardel, C., S. Benson, J. Hunt, S. Michaelis and J. Beckwith. 1987. J. Bactriol. 169:1286-1290.
8. Ito, K., M. Wittekind, M. Nomura, K. Shiba, T. Yura, A. Miura and H. Nashimoto. 1983. Cell 32:789-797.
9. Fandl, J.P. and P.C. Tai. 1987. Proc. Natl. Acad. Sci. USA 84:7448-7452.

10. Crooke, E. and W.T. Wickner. 1987. Proc. Natl. Acad. Sci. USA 84:5216-5220.
11. Weiss, J.B., P.H. Ray, and P.J. Bassford. 1988. Proc. Natl. Acad. Sci. USA 85:8978-8982.
12. Eilers, M. and G. Schatz. 1988. Cell 52:481-483.
13. Minsky, A., R.G. Summers RG and J.R. Knowles. 1986. Proc. Natl. Aca. Sci. 83:4180-4184.
14. Fitts, R., Reuveny, J. van Amsterdam, J. Mulholland and D. Botstein. 1987. Proc. Natl. Acad. Sci. USA 84:8540-8543.
15. Pohler, J., H. Halter, K. Beyreuther and T.F. Meyer. 1987. Nature (London) 325:458-462.
16. Wong, K.R., M.J. Green and J.T. Buckley. 1984. J. Bacteriol. 171:2523-2527.
17. Hirst, T.R., L.L. Randall and S.J.S. Hardy. 1984. J. Bacteriol. 157:637-642.
18. Hirst, T.R. and J. Holmgren. 1987. J. Bacteriol. 169: 1037-1045.
19. Gray, G.L., D.H. Smith, J.S. Baldridge, R.N. Harkins, M.L. Vasil, E.Y. Chen and H.L. Heynedker. 1984. Proc. Natl. Acad. Sci. USA 81:2645-2649.
20. Holmgren J. 1981. Nature (London) 292:413-417.
21. Mekalanos J.J., D.J. Swartz, G.D.N. Pearson, N. Harford, F. Groyne and M. de Wilde. 1983. Nature (London) 306: 551-557.
22. Yamamoto, T., T. Tamura and T. Yokota. 1984. J. Biol. Chem. 259:5037-5044.
23. Leong, J., A.C. Vinal and W.S. Dallas. 1985. Infect. Immun. 48:73-77.
24. Ito, K. and J.R. Beckwith. 1981. Cell 25:143-150.
25. Strauch, K.L. and J. Beckwith. 1988. Proc. Natl. Acad. Sci. USA 85:1570-1580.
26. Hirst, T.R., J. Sanchez, J.B. Kaper, S.J.S. Hardy and J. Holmgren. 1984. Proc. Natl. Acad. Sci. USA 81:7752-7756.
27. Hardy, S.J.S., J. Holmgren, S. Johanasson, J. Sanchez and T.R. Hirst. 1988. Proc. Natl. Acad. Sci. USA 85: 71009-710013.
28. Sandkvist, M., T.R. Hirst and M. Bagdasarian. 1987. J. Bacteriol. 169:4570-4576.
29. Hirst, T.R. and J. Holmgren. 1987. Proc. Natl. Acad. Sci. USA 84:7418-7422.

GENETICS OF *YOP* PRODUCTION IN *YERSINIA ENTEROCOLITICA*

Guy Cornelis, T. Biot, C. Lambert de Rouvroit, T. Michiels, B. Mulder, C. Sluiters, M.-P. Sory, M. Van Bouchaute, J.-C. Vanooteghem and P. Wattiaux

Unite' de Microbiologie, Université Catholique de Louvain, UCL/5490 - B 1200 Brussels, Belgium

INTRODUCTION

Y. enterocolitica is a common human pathogen which causes gastrointestinal syndromes of various severities, ranging from mild diarrhoea to mesenteric adenitis evoking an appendicitis, systemic involvement is unusual with *Y. enterocolitica* but arthritis and erythema nodosum are common complications.

Y. enterocolitica shares with *Y. pseudotuberculosis* and *Y. pestis* a marked tropism for lymphoid tissue and a remarkable ability to resist the primary immune response of the host (for a complete review see ref. 1).

The three virulent yersiniae release large amounts of a set of proteins called Yops and encoded by a 70kb plasmid. The loss of this property always correlates with the loss of pathogenicity. This paper is devoted to the genetics of the production of these Yop proteins.

PRODUCTION OF YOP PROTEINS

Calcium dependency

When incubated at 37°C, in the absence of calcium ions, virulent yersiniae restrict their growth and synthesize large amounts of Yops. This growth restriction phenomenon, associated with Yops production, is generally known as Ca^{++} dependency. Although, it is quite obvious that massive production and secretion of a given set of proteins - up to 10% of the total bacterial protein mass (Michiels et al., in preparation) must divert the metabolic potential from growth, the phenotype of some mutants suggest that a specific factor could be involved in the repression of growth at 37°C, in the absence of Ca^{++} (see below).

Status of Yops

The Yop symbol - for *Yersinia* Outer membrane Protein -

Microbial Surface Components and Toxins in Relation to Pathogenesis
Edited by E.Z. Ron and S. Rottem, Plenum Press, New York, 1991

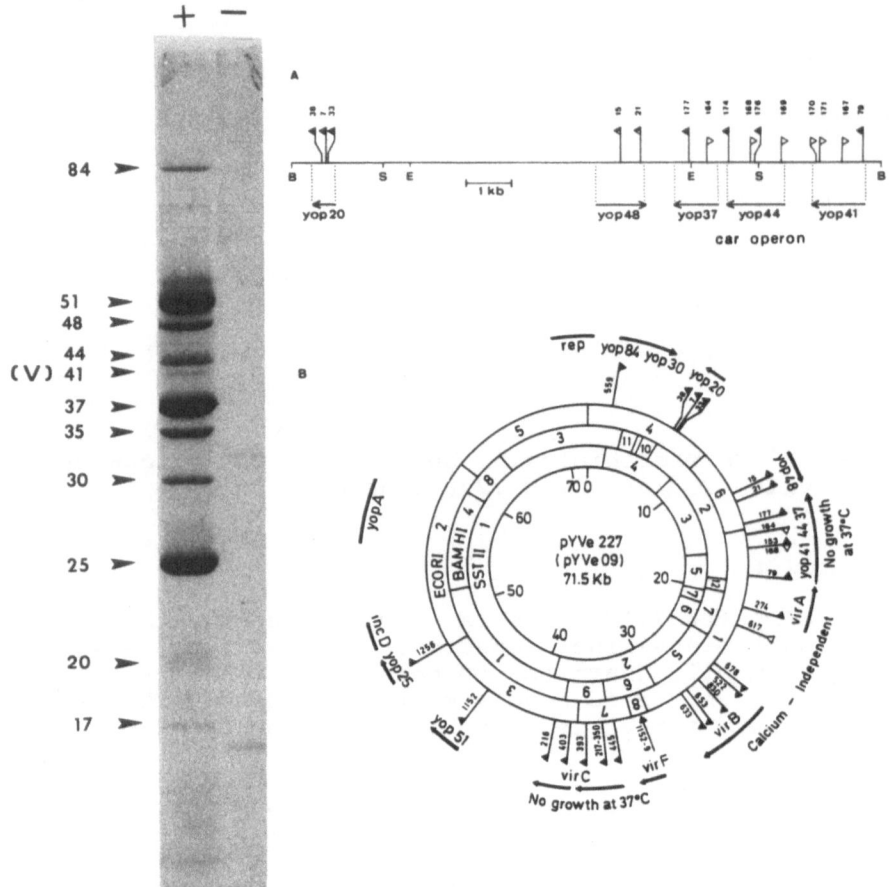

Fig. 1.   Left: SDS-PAGE analysis of the Yops released in the
culture medium of *Y. enterocolitica* W227(pYVe227)
upon thermal induction (+). The plasmid-less variant (-) does not secrete Yops. Yops are identified
by their molecular mass (in kilodaltons). Note that
Yop41 is the V antigen.
Right: Genetic map of pYVe227, a typical pYV plasmid from a *Y. enterocolitica* strain of serotype
O:9. *rep* is the replication region; *incD* is the
stabilization and incompatibility region; *vir* are
genes involved in the growth restriction and release of the Yops. Gene *virF* encodes the transcription activator controlling *yop* genes. *yop* genes are
genes encoding the released proteins. The numbers
refer to the molecular mass of the proteins. Gene
yopA encodes protein P1, which forms a fibrillar
matrix on the surface of *Y. enterocolitica* . Genes
yop41, yop44, and *yop37* are part of the *car* operon. The arrows point in the direction of transcription. The flags and the small numbers identify the
insertion mutation that defined the genes. The data
are from Balligand et al (19); Cornelis et al.
(1,8,9) and Biot and Cornelis (27).
Part A: detail of *BamHI* fragment 2 showing the *car*
operon.
Part B: total map of pYVe227.

originates from the fact that these proteins were first detected in outer membrane preparations of *Y. enterocolitica* and *Y. pseudotuberculosis* (2,3). However, in *Y. enterocolitica*, Yop proteins are released in the culture supernatant (4) (Figure 1) and their status of outer membrane proteins is questionable (5). One of the proteins released by *Y. enterocolitica* also appears in the cytoplasm of cells in growth-restricted cultures. This protein corresponds to the V antigen, a virulence associated factor discovered in *Y. pestis* in the mid fifties (6).

The set of Yops

The set of Yops is highly conserved in *Y. enterocolitica*, *Y. pseudotuberculosis* and *Y. pestis*: they are immunologically related and most of the proteins have similar molecular masses (5,7). However, it is still hazardous to establish the complete correspondence table between the Yops produced by the three species, Nevertheless, taking the genetic data into account, the match can be established for the major Yops and it appears that the species variability in the repertory of Yops is limited to a few proteins. *Y. enterocolitica* W227 releases at least 11 Yops (8,9, Mulder et al., in preparation), identified by their apparent molecular mass (Figure 1).

Role of the Yops in pathogenesis

Although Yops production is generally considered as essential for virulence, there is at present only scarce information about their individual role. The loss of some of the Yops has been shown, in the three species, to lead to a severe increase in the LD_{50} for the mouse (10-13, Mulder et al,m in preparation). Rosqvist et al. (14) showed that Yop2b from *Y. pseudotuberculosis* inhibits phagocytosis by mouse peritoneal macrophages cultured *in vitro*.

The adhesin P1

Y. enterocolitica synthesize a large surface associated protein which we call P1 in this paper. In contrast to the release of Yops, production of this protein does not correlate with growth restriction. Protein P1 consists of subunits of about 50 kDa (15) and forms a fibrillar matrix covering the outer membrane (16,17). This protein is regarded as an adhesin. At least in *Y. enterocolitica*, it was shown to mediate adherence of human epithelial cells (18), to contribute to the intestine colonization (16) and to confer resistance to the bactericidal activity of human serum (19).

GENETICS OF YOPs PRODUCTION

The pYVe plasmid and *yop* genes of *Y. enterocolitica*

Yop proteins production is governed by a 70kb plasmid called pYVe for "plasmid involved in Yersinial Virulence". Nine genes encoding Yops were mapped on pYVe227 from *Y. enterocolitica* W227 by insertion mutagenesis with a mini-Mu*dlac* or *Tn2507*, a transposon containing a chloramphenicol acetyltransferase gene devoid of its promoter (8,9, Mulder et al., in preparation). Five *yop* genes are arranged in two divergently transcribed operons. One of these operons encodes Yop37, Yop44 and Yop41, the V antigen. Any insertion in this operon of pYVe227 makes the *Y. enterocolitica* host strain unable to grow at 37°C, even in the presence of Ca++, however, these mutants release all the Yops (apart from those encoded by genes downstream from the insertion, which are not produced). The distal mutations in this operon make the host strain not only unable to grow at 37°C but even sensitive to Ca²+ at this temperature (Mulder et al., in preparation). These particular phenotypes remain unaccounted for but they suggest that (i) a specific factor could be involved in the restriction of growth and (ii) that this operon is somehow involved in the Ca²+ response. Hence, we called it the *car* operon (for Ca²+ regulation).

The second operon contains *yop30*. Four other *yop* genes are scattered around pYVe (Figure 1). Mutations in these genes result in the loss of the corresponding Yop without alteration of the Ca²+ dependency phenotype (8,9, Mulder et al., in preparation).

Transcription of the *yop* genes was measured by assaying β-galactosidase or chloramphenicol acetyltransferase produced by the mutants containing *yop-lacZ* or *yop-cat* operon fusions. for most *yop* genes, transcription is stimulated more than 100-fold after a temperature shift from 28°C to 37°C. Surprisingly, Ca²+ only reduces transcription by a factor of 1 to 7. since Ca²+ prevents the apparition of the Yops, we infer that it primarily acts at a post-transcriptional stage, I.E., translation or export (9).

Cloned *yop* genes complement the corresponding *yop* mutations on the pYVe plasmid but they do not direct the synthesis of their product when they are introduced in pYVe - *Y. enterocolitica* strains. This indicates that pYVe genes, other than *yop* genes, are required for the production of Yops.

vir genes of *Y. enterocolitica*

Mutations in a 17kb region of pYVe abolish Ca²+ dependency, the release of the entire set of Yops and virulence. Our genetic analysis of this region of pYVe227 allowed us to identify at least four transcription units called *virA, B, C*, and *F* (8,9,20). Transcription of these genes is stimulated by temperature and, as for *yop* genes, is poorly affected by Ca²+. Mutants affected in *virC* grow poorly at 37°C, even in the presence of Ca²+ and they do not produce Yops. The region encompassing the *vir* genes is referred to as the "Calcium region" (Figure 1).

In order to evaluate the role of the various *vir* loci on transcription of the *yop* genes, we introduced a hybrid *yop51-cat* operon into *Y. enterocolitica* strains carrying various *vir* mutations. This operon appeared to be transcribed in pYVe+ derivatives of *Y. enterocolitica* but not in pYVe-variants. Transcription of the *yop51-cat* unit was reduced in *virA, B* or *C* mutants and completely abolished in the *virF* mutant. Gene *virF* corresponds to gene *lcrF* of *Y. pestis* KIM, previously identified by Yother et al. (21).

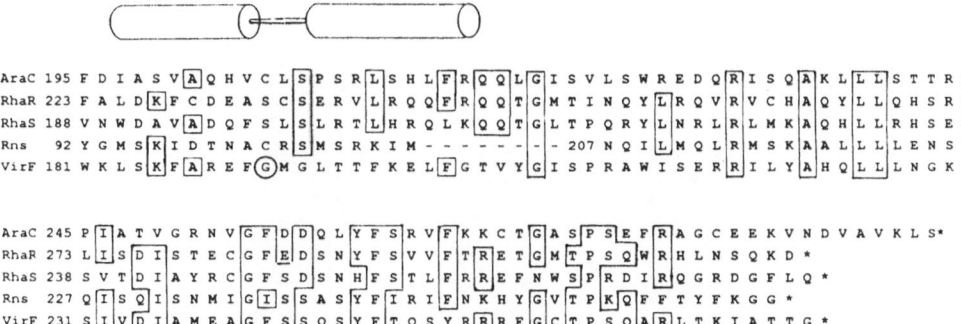

```
Helix 2              Helix 3

AraC 195 F D I A S V Ⓐ Q H V C L Ⓢ P S R Ⓛ S H L Ⓕ R Ⓠ Ⓠ Ⓛ Ⓖ I S V L S W R E D Ⓠ Ⓡ I S Q Ⓐ K L Ⓛ Ⓛ S T T R M
RhaR 223 F A L D Ⓚ Ⓕ C D E A S Ⓒ Ⓢ E R V Ⓛ R Q Ⓠ Ⓕ R Ⓠ Ⓠ Ⓣ Ⓖ M T I N Q Y Ⓛ R Q V R V C H A Ⓠ Y Ⓛ Ⓛ Q H S R L
RhaS 188 V N W D A V Ⓐ D Q F S L Ⓢ Ⓛ R T Ⓛ H R Q L K Ⓠ Ⓠ Ⓣ Ⓖ L T P Q R Y Ⓛ N R L Ⓡ L M K A Ⓠ H Ⓛ Ⓛ R H S E A
Rns   92 Y G M S Ⓚ I D T N A C R Ⓢ M S R K I M - - - - - - - 207 N Q I Ⓛ M Q L Ⓡ M S K A A Ⓛ Ⓛ L E N S Y
VirF 181 W K L S Ⓚ Ⓕ A R E F Ⓖ M G L T T F K E L Ⓕ G T V Y Ⓖ I S P R A W I S E R Ⓡ I L Y A Ⓗ Q Ⓛ Ⓛ L N G K M
```

```
AraC 245 P Ⓘ A T V G R N V Ⓖ F Ⓓ Ⓓ Q L Ⓨ F Ⓢ R V Ⓕ K K C T Ⓖ A Ⓢ P Ⓢ E F Ⓡ A G C E E K V N D V A V K L S *
RhaR 273 L Ⓘ S Ⓓ I S T E C G F Ⓔ D S N Ⓨ F S V V F T R Ⓔ T G M Ⓣ P S Q W R H L N S Q K D *
RhaS 238 S V T Ⓓ I A Y R C G F S Ⓓ S N H F S T L Ⓕ R Ⓡ E F N W S P R D I Ⓡ Q G R D G F L Q *
Rns   227 Q Ⓘ S Ⓠ I S N M I Ⓖ I Ⓢ S A S Ⓨ F Ⓘ R I Ⓕ N K H Y G V Ⓣ P K Ⓠ F F T Y F K G G *
VirF 231 S I Ⓥ Ⓓ I A M E A G F S S Q S Ⓨ F T Q S Y R Ⓡ R F G C Ⓣ P S Q A Ⓡ L T K I A T T G *
```

Fig. 2. Amino acid sequence homology of VirF to the regulators of the arabinose (22) rhamnose (23) and the CD pili (24). Only amino acids found in at least 3 proteins are boxed. Note that a break was introduced in the sequence of Rns for the alignment. The glycine residue encircled is conserved in the great majority of cro-like proteins.

VirF, the *yop* regulator

Gene *virF* is thus required for the transcription of *yop* genes. *VirF* is also sufficient to activate *yop* genes transcription since a *Y. enterocolitica* strain carrying only the *yop51-cat* unit and *virF* produces chloramphenicol acetyltransferase (20). Gene *virF* is thus a positive regulator of transcription of the *yop* genes and the *yop* genes constitute a regulon. Gene *virF* is itself transcribed in a thermodependent manner which explains the thermoregulation of transcription of *yop* genes. Ca²⁺ has little influence on transcription of *virF* (20). Whether *virF* is itself autoregulated or negatively regulated by a *Yersinia* chromosomal gene is still unclear.

VirF is a 30,879 daltons protein. Its carboxy-terminal half shares significant amino acid sequence homology with four other transcription regulators. These are AraC, the arabinose operon regulator (22), RhaS and RhaR, regulators of the rhamnose operon (23) and the *E. coli* regulator of CS pili (24) (Figure 2). The homologous region includes the Cro-like DNA binding domain of AraC and the sequence of this domain of VirF is perfectly compatible with an helix-turn-helix conformation (Figure 2).

Gene *virF* was cloned downstream a T7 promoter and over-produced upon induction of the synthesis of T7 RNA polymerase (25). The cells were harvested, disrupted by sonication and VirF was purified by ultracentrifugation, gel filtration and phosphocellulose ion-exchange chromatography. Gel filtration experiments suggested that VirF presumably exists as a dimer.

Mobility shifts assays indicates that VirF binds to DNA fragments containing sequences upstream *yop51* and *yop25* (Figure 3). DNase I cleavage protection experiments identified a 31 nucleotides protected sequence upstream the *yop51* promoter (Figure 4). There is no clear consensus sequence emerging from the comparison of these two sequences. The

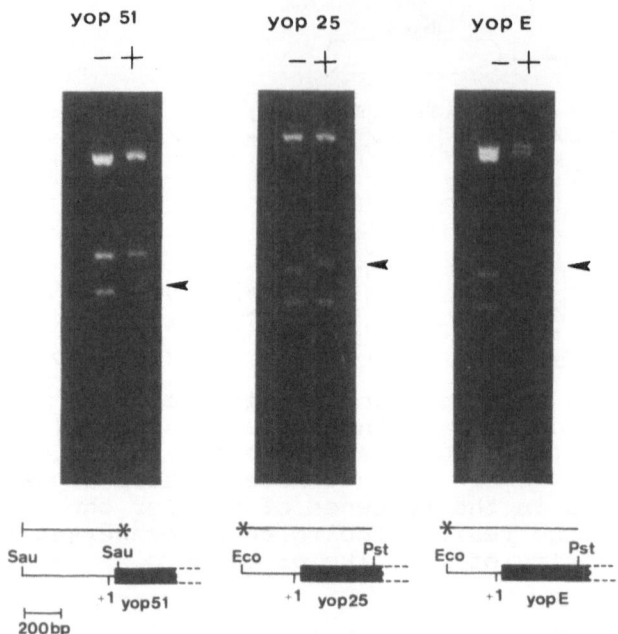

Fig. 3.   Gel mobility shift analysis of VirF binding to the
          yop promoter region. Restriction fragments of plas-
          mids (about 200 ng) were incubated with no protein
          (-) or with VirF at an approximate concentration of
          80 nm (+). The binding mixture was 10 mM Tris-HCl
          pH 7.4; 1 mM EDTA; 5% glycerol and 50 mM KCl. After
          a 15 minutes binding at 37°C, the samples were
          electrophoresed on a 1.2% agarose gel in TAE buffer
          and stained with ethidium bromide. The arrows indi-
          cate the *yop* promoter containing fragments. The
          other fragments are either vector or pYV DNA devoid
          of any *yop* promoter.

definition of such a sequence will await the identification
of more *yopI* sites (Lambert de Rouvroit et al., in preparation).

Regulation of *vir* genes

     We cloned the central part of *virF* on a suicide vector
derived from pJM703 (26) and the recombinant plasmid served
to mutate *virF* by homologous recombination. we mutated *virF*
on pYVe plasmids carrying either a *virA-lacZ* fusion or a
virB-lacZ fusion and we measured the β-galactosidase activity
at 28°C and 37°C. This activity was found to be of the same
order for the *virF+* and *virF-* isogenic strains, indicating
that transcription of *virA* and *virB* is not regulated by *virF*.
Since transcription of these genes is thermoregulated, one
must postulate the existence of another regulator.

PERSPECTIVES AND CONCLUSIONS

     The organization of *yop* genes as a regulon controlled by
a transcription activator explains the coordinate production

```
  VirF binding                    -35                        -10              yop51 mRNA

CCTATTATTAAAATAATACGACTAGCATTATAAGAAAAAATATTTTTTATGTTTATAGTATAGGCGTGTATTTAATTAG
GGATAATAATTTTATTATGCTGATCGTAATATTCTTTTTTATAAAAAAATACAAATATCATATCCGCACATAAATTAATC
      *         *         *         *         *         *         *         *
    -60       -50       -40       -30       -20       -10        +1
```

Fig. 4. Nucleotide sequence of the promoter region of *yop51* showing the start of transcription, the promoter and the VirF binding region. The latter region was determined on both strands by DNaseI footprinting.

of the set of Yop proteins. The fact that transcription of the activator *virF* itself is thermodependent accounts for the effect of temperature on transcription of the *yop* gene but the mechanism of regulation of the *vir* genes, including *virF* itself, remains to be clarified. At this stage, the major challenge consists in the understanding of the role of Ca^{2+}.

Although it is clear that Ca^{2+} reduces transcription of the *yop* genes, there is at present no agreement as to the incidence of this effect. We consider that a transcription reduction of 3 fold - as observed for most *yop* genes - does not explain the disappearance of the Yops and we hypothesized that Ca^{2+} primarily acts at a post-transcriptional stage, i.e., translation or export. In this hypothesis, the observed transcription reduction in the presence of Ca^{2+} would simply result from a feedback inhibition resulting from the coupling of transcription, translation and export. Apart from its possible role in the export of the Yops, Ca^{2+} could also trigger a growth restriction mechanism. However, so far, we have been unable to identify a Ca^{2+} sensor gene in *Y. enterocolitica*.

SUMMARY

Growth of *Y. enterocolitica* is restricted at 37°C in the absence of calcium ions. This phenomenon correlates with the massive release of a set of 11 proteins called Yops. Growth restriction and Yops production are governed by a 70kb plasmid called pYVe. Nine *yop* genes were mapped on pYVe227, the natural plasmid from a typical serotype 9 strain. Genes *yop41* (encoding the V antigen), *yop44* and *yop37* form an operon. Mutants affected in this operon do not grow or even die at 37°C in the presence of Ca^{2+}. We concluded that this operon is somehow involved in the Ca$_{2+}$ response and we called it *car*. Genes *yop84* and *yop30* constitute a second operon. The four other *yop* genes mapped so far are scattered around pYVe. The *yop* genes and the *car* operon constitute a thermoactivated regulon controlled by the gene *virF*. The transcription activator VirF is a 30kDa protein that binds to a 28-30 nucleotide sequence upstream of *yop25* and *yop51*. It is a member of a new family of regulators including those of the arabinose and rhamnose operons as well as a regulator of enteric colonization pili. The production of the Yops also involves other *vir* genes. So far, we identified *virA* and *virB*. The product of gene *virF* is not required for transcription of *virA* and *virB*. The latter genes must thus be controlled by another regulator. The role of calcium ions on the release of Yops remains largely unknown.

ACKNOWLEDGEMENTS

Our work on *Y. enterocolitica* is supported by an "Action Concertee" (contract 86/91-86) from the Belgian Ministry of Sciences (SPPS) and by a grant from the Belgian Fund for Medical Research (FRSM) (contract 3.4514.83). T. Michiels and C. Lambert de Rouvroit are fellows of the Belgian National Fund for Scientific Research (FNRS). T. Biot, B. Mulder, M.-P. Sory and J.-C. Vanooteghem are fellows of the Belgian Institute for Scientific Research applied to Industry and Agriculture (IRSIA).

REFERENCES

1. Cornelis, G., Y. Laricge, G. Balligand, M.-P. Sory and G. Wauters. 1978a. Rev. Infect. Dis. 9:64-87.
2. Portnoy, D.A., S.L. Moseley and S. Falkow. 1981. Infec. Immun. 31:775-782.
3. Straley, S.C. and R.R. Brubaker. 1981. Proc. Natl, Acad, Sci. USA. 78:1224-1228.
4. Heesemann, J., B. Algermissen and R. Laufs. 1984. Infect. Immun. 46:105-110.
5. Heesemann, J. U. Gross, N. Schmidt and R. Laufs. 1986. Infect. Immun. 54:561-567.
6. Burrows, T.W. and G.A. Bacon. 1956. Br. J. Exp. Pathol. 37:481-493.
7. Bolin,I., D.A. Portnoy and H. Wolf-Watz. 1985. Infect. Immun. 48:234-240.
8. Cornelis, G., M.-P. Sory, Y. Laroch and I. Derclays. 1986. Microb. Pathogen. 1:349-359.
9. Cornelis, G., J.-C. Vanooteghem and C. Sluiters. 1987b. Microb. Pathogen. 2:367-379.
10. Bolin,I. and H. Wolf-Watz. 1988. Mol. Microbiol. 2:237-245.
11. Forsberg, A. and H. Wolf-Watz. 1988. Mol. Microbiol. 2:121-133.
12. Starley, S.C. and W.S. Bowmer. 1988. Infect. Immun. 51:445-454.
13. Sory, M.-P. and G. Cornelis. 1988. Microb. Pathogen. 4:431-442.
14. Rosqvist, R., I. Bolin and H. Wolf-Watz. 1988. Infect. Immun. 56:2139-2143.
15. Skurnik, M. and H. Wolf-Watz. 1989. Mol. Microbiol. 3:517-529.
16. Kapperud, G., E. Namork, M. Skurnik and T. Nesbakken. 1987. Infect. Immun. 55:2247-2254.
17. Zaleska, M., K. Lounatmaa, M. Nurmine, E. Wahlstrom and H. Makela. 1985. EMBO J. 4:1013-1018.
18. Heesemann, J. and L. Gruter. 1987. FEMS Microbiol. Letters 40:37-41.
19. Balligand,G., Y. Laroche and G. Cornelis. 1985. Infect. Immun. 48:782-786.
20. Cornelis,G., C. Sluiters, C. Lambert de Rouvroit and T. Michiels. 1989. J. Bacteriol. 171:254-262.
21. Yother,J., T.W. Chamness and J.D. Goguen. 1986. J. Bacteriol. 165:443-447.
22. Wallace, R.G., N. Lee and A.V. Fowler. 1980. Gene 12:179-190.
23. Tobin, J.F. and R.F. Schleif. 1987. J. Mol. Biol. 196:789-799.

24. Caron, J., L.M. Coffield and J.R. Scott. 1989. Proc. Natl. Acad. Sci. USA 86:963-967.
25. Tabor, S. and C.C. Richardson. 1985. Proc. Natl. Acad. Sci. USA 82:1074-1078.
26. Miller, V.L. and J.J. Mekalanos. 1988. J. Bacteriol. 170:2575-2583.
27. Biot, T. and G. Cornelis. 1988. J. Gen. Microbiol. 134:1525-1534.

TNF-α, IL-1 AND PGE$_2$ SECRETION FOLLOWING MACROPHAGE

ACTIVATION BY *MYCOPLASMA CAPRICOLUM* MEMBRANES

T. Sher[1] A. Yamin[1], I. Stein[1],
S. Rottem[2] and R. Gallily[1]

[1]The Lautenberg Center for General and Tumor Immunology
and   [2]The Department of Ultrastructure Research, The
Hebrew University-Hadassah Medical School.P.O.Box 1172
Jerusalem, Israel 91010

INTRODUCTION

Macrophages play an important role in the body's defence against microbial infections and neoplasia. Activation of macrophage to selectively kill tumor cells is the result of interaction with various microbial products and cytokines (1,2,3). Several cytotoxic factors secreted from activated macrophage have been described, the most important among them is tumor necrosis factor α (TNF-α) (5). Other factors implicated in macrophage-mediated tumor cell cytotoxicity are: interleukin-1 (IL-1) (6), cytolytic serine proteases, reactive oxygen products, arginase, lysosomal enzymes and the complement factor C3a (7). TNF-α causes hemorrhagic necrosis when injected into tumor bearing mice and is also toxic to certain tumor cell lines *in vitro*. The common potent inducer of TNF-α is bacterial LPS (6). IL-1 apart from contributing to macrophage-mediated cytolysis, is a major regulator of T lymphocyte proliferation (8). Macrophage were shown to remain only transiently activated, such that the induced tumor cytotoxicity is rapidly lost (9,10). This led to the suggestion that a negative feedback regulation may exist and that prostaglandin E$_2$ (PGE$_2$) may represent an endogenous regulator that shuts off macrophage activation once it has developed (11). Since LPS ,the common inducer of TNF-α and IL-1, is highly toxic in human and animals (5,12), attempts are directed at finding a non-toxic substitute. We recently found that membranes of *Mycoplasma capricolum* , a non-toxic mycoplasma which contain no LPS (13), induce TNF-α-mediated tumor cell killing following murine bone marrow macrophage (BM-macrophage) activation (14,15). In the work presented herein, we describe the potency of *M. capricolum* membranes to induce TNF-α, IL-1 and PGE$_2$ secretion following BM-macrophage activation. We discuss the exerted effects of *M. capricolum* membranes on the immune system and their therapeutic potential in treatment of malignant diseases.

Microbial Surface Components and Toxins in Relation to Pathogenesis
Edited by E.Z. Ron and S. Rottem, Plenum Press, New York, 1991

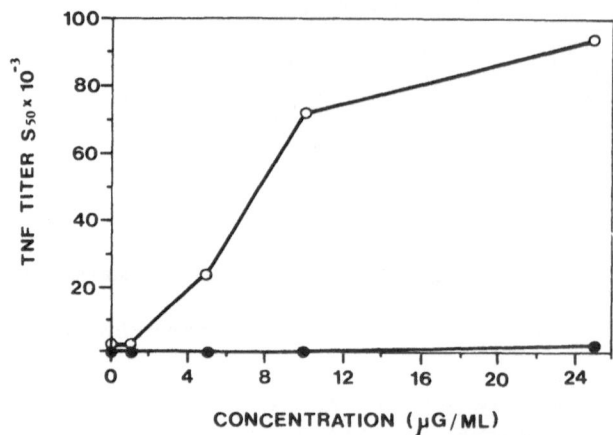

Fig. 1.   TNF-α titers secreted by C3H/HeJ BM-macrophage
          following activation by either *M. capricolum* mem-
          branes (o) or LPS (o); a dose response curve. BM-
          macrophage layered in 96 microwell plates were
          incubated with *M. capricolum* membranes (0.1-25 μg
          protein/ml) or LPS (0.1-25 μg/ml) in DMEM medium.
          Twenty four hours later three-fold dilutions of the
          supernatants were added to BALB/c CL.7 cells in the
          presence of actinomycin D. The plates were stained
          after 20 hours and TNF-α titers were calculated.

INDUCTION OF TNF-α SECRETION BY *M. capricolum* MEMBRANES IN
LPS-LOW RESPONSIVE C3H/HeJ-DERIVED BM-MACROPHAGE

        Figure 1 shows a dose-response curve of TNF-α titers
secreted by BM-macrophage from C3H/HeJ mice, a strain with a
defective LPS response gene *Lps* (24), following activation by
various concentrations of either *M. capricolum* membranes or
LPS. *M. capricolum* membranes induced high-titer TNF-α secretion, as compared to the negligible levels induced by LPS.

IL-1 SECRETION FOLLOWING C57BL/6-BM-MACROPHAGE ACTIVATION BY
M. CAPRICOLUM MEMBRANES

        The IL-1 content was determined in a thymocyte-prolife-
ration assay. Figure 2 shows the secretion of IL-1 by C57BL/-
6-derived BM-macrophage in response to incubation with various concentrations (1,10 and 25 μg protein/ml) of mycoplasma
membranes. The highest level of IL-1 release obtained following activation by 10 μg protein/ml of *M. capricolum* membranes
was similar to that secreted in response to activation by 10
μg/ml LPS. Under the experimental conditions, the amount of
spontaneously secreted IL-1 was negligible.

IL-1 SECRETION BY BM-MACROPHAGE   FROM LPS-LOW-RESPONSIVE
C3H/HeJ MICE

        It was shown above that *M. capricolum* membranes acti-
vated C3H/HeJ-BM-macrophage to high-level TNF-α secretion
whereas these cells hardly responded to activation by LPS.
Figure 3 shows that a similar response is observed with

202

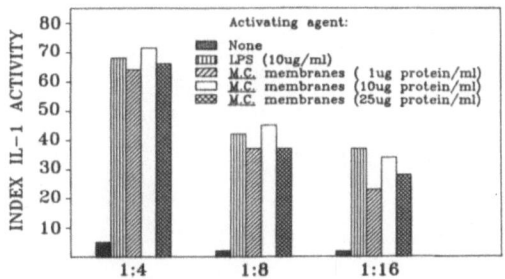

DILUTION OF TESTED SUPERNATANTS

Fig. 2.   IL-1 activity determination following C57BL/6 BM-
            macrophage activation by various concentrations of
            M. capricolum membranes (1,10,25 µg protein/ml) or
            LPS (10 µg/ml) as designated above in the figure,
            in the presence of indomethacin (5x10⁵). Twenty
            four hour-supernatants were collected and assayed
            for IL-1 activity at 1:4, 1:8 and 1:16 dilutions.
            Results are expressed as index of IL-1 activity.

regards to IL-1 secretion. The level of IL-1 activity detected following BM-macrophage activation by *M. capricolum* membranes (10 µg protein/ml) was about 3 times higher than that
obtained following activation by LPS (10µg/ml). The IL-1
activity exhibited by C3H/HeJ (H-2ᵏ) BM-macrophages following
mycoplasma-membrane activation was quite similar to that
exhibited by C3H/CrgI BM-macrophages, a strain of mice having
normal *Lps* gene and the same H-2 as C3H/HeJ mice. In accordance with our findings concerning C57BL/6 BM-macrophages
(Figure 2), both mycoplasma membranes (10µg protein/ml) and
LPS (10µg/ml) induced similar level of IL-1 secretion by
C3H/CrgI mice.

PGE₂ AND TNF-α INDUCTION BY *M. CAPRICOLUM* MEMBRANES AND LPS

The mutual control relationship between TNF-α and PGE₂
(11,25) led us to examine their relative amounts secreted by
BM-macrophages following incubation with *M. capricolum* membranes or LPS. As shown in Table 1, increasing concentrations
of *M. capricolum* membranes from 1 to 25 µg protein/ml, resulted in 365 times augmentation in TNF-α titers while PGE₂
concentration was increased 8.3 folds only. Thus the ratio
TNF-α/PGE₂ is much higher following activation by mycoplasma
membranes compared to activation by LPS. Low amounts of PGE₂
were spontaneously secreted from untreated BM-macrophages
while no spontaneous secretion of TNF-α was found.

Our studies show that *M. capricolum* membranes, like LPS,
are potent inducers of TNF-α and IL-1 secretion by activated
BM-macrophages. As *M.capricolum* membranes lack LPS (13) it is
assumed that the mechanism of macrophage activation by *M.
capricolum* differs from that of LPS. Supporting this hypothesis are the results obtained with BM-macrophages derived from
LPS low-responsive-C3H/HeJ mice. Such macrophage when activated with *M. capricolum* membranes secreted high titers of
TNF-α compared to negligible titers induced by LPS. Beutler
et al. (26) suggested that a dual defect prevents TNF-α
expression in C3H/HeJ mice; namely, a diminished mRNA express

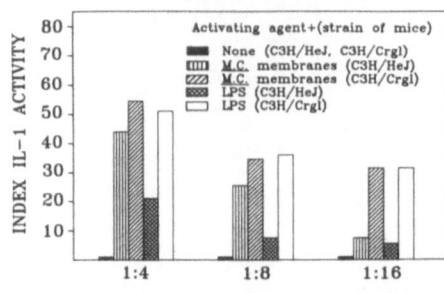

DILUTION OF TESTED SUPERNATANTS

Fig. 3. IL-1 activity determination following activation of BM-macrophages from either C3H/Crgl or LPS-low-responsive C3H/HeJ mice. Activation was carried out in the presence of indomethacin by either *M. capricolum* membranes (10 μg protein/ml) or LPS (10 μg/ml). Assay was performed and Index of IL-1 activity was calculated as described in Figure 2.

ion in response to low concentrations of LPS, and a post--transcriptional defect inhibiting TNF-α protein production. It appears that the mechanism by which M. capricolum membranes circumvents the blockage imposed by the defective *Lps* gene renders the C3H/HeJ BM-macrophages capable of producing high levels of TNF-α. The increase in IL-1 secretion by C3H/HeJ BM-macrophages following *M. capricolum* membranes compared to activation by LPS is less pronounced then that observed for secretion of TNF-α. Our results showing that mycoplasma membranes are capable of activating macrophagess for IL-1 release are compatible with previous findings concerning another mycoplasma, *M. arthritidis*. The mitogenic effect on murine and human T lymphocytes originally attributed to direct effect of *M. arthritidis* supernatants (MAS) (27), has recently been revised and shown to be partially due to the presence of IL-1 in the assay system. This IL-1 was secreted by accessory cells, probably macrophages, following activation by MAS (28). In addition to induction of TNF-α and IL-1 secretion by activated BM-macrophagess, we have presented for the first time evidence that mycoplasma membranes induce PGE$_2$ release. Thus, together with our previous observation on their mitogenic activity (15), it seems that mycoplasma membranes have a broad effect on the immune system. TNF-α and IL-1, apart from participating in the macrophages--mediated tumor cell lysis *in vitro,* have many other functions (5,8,29). PGE$_2$ has been reported to be involved in regulation of a variety of cellular and humoral immune responses , including inflammation. For example, it acts as a feedback inhibitor of T-cell activation, and modulate B-cell activation and antibody production (30). The effect of *M. capricolum* membranes is being further complicated considering the inter-relationship between TNF-α, IL-1 PGE$_2$ and TNF-α induces IL-1 production (31). TNF-α and IL-1 synergize to highly augment PGE$_2$ secretion (32).PGE$_2$ in turn, was shown to have two opposite, dose-dependent effects on TNF-α release from macrophages : low PGE$_2$ concentrations (0.1 to 10 ng/ml) stimulated, whereas higher concentrations suppressed TNF-α release (33). In our experiments we simultaneously measured TNF-α titers and PGE$_2$ concentrations following BM-macro-phagess activation by either *M. capricolum* membranes or LPS.

Table 1. Relative secretion of TNF-α and PGE$_2$ following activation by M. capricolum membranes and LPS.

Order No.	Activating agent	Concentration of activating agent	TNF-α titer S_{50}	PGE$_2$ pg/ml
		(μg/ml)		
1	LPS	1	29,340	1,290
2	LPS	5	7,150	910
3	LPS	25	12,910	1,190
		(μg protein/ml)		
4	*M.C. membranes	1	460	250
5	*M.C. membranes	5	9,340	820
6	*M.C. membranes	25	168,340	2,080
7	Medium (control)	–	0	50

*M.C. = Mycoplasma capricolum.

As higher amounts of TNF-α relative to PGE$_2$ were secreted following macrophages activation by mycoplasma membranes, a much higher TNF-α/PGE$_2$ ratio was obtained compared to that observed in LPS-activated BM-macrophagess. This may account in part, for the high in vivo toxicity of LPS as compared to the non-toxic nature of M. capricolum. Indeed, in a preliminary assay carried out with BCG primed-C57BL/6 mice, injection of M. capricolum membranes (up to 1000 μg protein/ml) were non-toxic (data not shown). Our results are compatible with reports on the involvement of PGE$_2$ in endotoxin induced shock. Combined administration of TNF-α or LPS with indomethacin, inhibitor of PGE$_2$ production, reduced lethality without blocking the anti-tumor activity of TNF-α (34). The inevitable conclusion drawn from our experiments is that due to the broad effects M. capricolum membranes exert on the immune system their therapeutic potential in cancer treatment, should be evaluated by in vivo studies in tumor bearing mice.

SUMMARY

We have previously reported that membranes of Mycoplasma capricolum activate murine bone marrow (BM) macrophages to tumor necrosis factor (TNF-α)-mediated tumor cell killing. We show herein, that M. capricolum membranes are potent inducers of IL-1 secretion, as well. M. capricolum membranes activate LPS-low-responsive C3H/HeJ-derived BM-macrophagess to high titer TNF-α secretion in contrast to negligible effect obtained following activation by LPS. IL-1 secretion in response to similar activation is also increased, though to a lesser extent. Both findings suggest that the mechanism of BM-macrophagess activation operated by M. capricolum membranes is different from that of LPS. Concomitantly with the induction of TNF-α secretion by *M. capricolum* membranes activated-BM-macrophagess, PGE₂ is released. However, the TNF-α/PGE₂ ratio obtained is much higher then that observed following LPS activation. The possibility that this difference bear relevance to the non-toxic nature of the mycoplasma membranes compared to the high toxicity of LPS, and the therapeutic potential of *M. capricolum* membranes in treatment of malignant diseases are discussed.

REFERENCES

1. Piessens, W.F., W.H. Churchill, Jr. and R. David. 1975. J. Immunol., 114:293-299.
2. Unanue, E.R. and P.M. Aller. 1987. Science, 236:551-557.
3. Boraschi, D. and A. Taagliabue. 1984. Lymphokines, 9:71-108.
4. Lohmann-Matthes, M.L., H. Lang, D. Sun, E. Kniepand B. Kickhofen. 1981. Lymphokine Reports 3,pp. 365-380, Academic Press.
5. Old, L.J.. 1985. Science, 230:630-632.
6. Onozaki, K., K. Matsushima, B.B. Aggarwal and J.J. Oppenheim. 1985. J. Immunol. 135:3962-3968.
7. Adams, D.O. and C.F. Nathan. 1983. Immunol. Today 4:166-170.
8. Dinarello, C.A.. 1984. Rev. Infect. Dis. 6:51-95.
9. Poste, G. and R. Kirch. 1979. Cancer Res. 39:2582-2589.
10. Taffet, S.M., J.L. Pace and S.W. Russel. 1981. J. Immunol. 127:121-124.
11. Schultz, R.M., N.A. Pavlidis, W.A. Stylos and M.A. Chirigas. 1987. Science, 202:320-321.
12. Bauss, F., W. Droge, and D.N. Mannel. 1987. Infect. Immun. 55:1622-1625.
13. Smith, P.F., T.A. Longworothy and W.R. Mayberry. 1976. J. Bacteriol., 125:916-922.
14. Sher, T., B. Uziely, Y. Ehrlich, A. YaminI. Stein, S. Rottem and R. Gallily. 1989. Isr. J. Med. Sci. 25:180.
15. Sher, T., A. Yamin, S. Rottem and R. Gallily. 1990. Zbl. Bakt. Hyg.. in press.
16. Johnson, W.J. and E. Balish. 1981. J. Reticuloendothel. Soc. 29:369-379.
17. Loewenstein, J. and R. Gallily. 1984. Tissue Culture and Res., pp. 395-405. Akademia Kiado, Publishing House of the Hungarian Acad. Sci.
18. Hayflick, L.. 1965. Texas Reports onBiology and Medicine, 23:285-303.
19. Razin, S. and S. Rottem. 1979. BiochemicalAnalysis of Membranes. pp. 3-26, Chapman andHall.
20. Lowry, O.H., N.J. Rosenbough, A.L. Farrand R.J. Randall. 1951. J. Biol. Chem. 193:265-275.
21. Ruff, M. and G.E. Gifford. 1981. Lymphokine Reports 2. pp. 235-257. Academic Press.
22. Treves, A.J., T. Tal, V. Barak and Z. Fuks. 1981. Eur. J. Immunol. 11:487-492.
23. Jaffer. B.M. and H.R. Behrman. 1974. Methods of Hormone Radioimmunoassay,pp. 19-26. Academic Press.
24. Ruco, L.P. and M.S. Meltzer. 1987. J. Immunol. 120:329-334.
25. Dayer, J.M., B. Beutler and A. Cerami. 1985. J. Exp. Med., 162:2163-2168.

26. Beutler, B., N. Krochin, I.W. Milsark, C.Luedke and A. Cerami. 1986. Science, 323:977-980.
27. Cole, B.C., G. Sullivan, R.A. Daynes,I.A. Sayed and J.R. Ward. 1982. J. Immunol. 128:2013-2018.
28. Bauer, A., M. Giese and H. Kirchner. 1989. Immunobiol. 179:124-130.
29. Cerami, A. and B. Beutler. 1988. Immunol. Today, 9:28-31.
30. Goodwin, J.S. and D.R. Webb. 1980. Clin. Immunol. Immunopath. 15:106-122.
31. Dinarello, C.A., J.G. Cannon, S.M. Wolff, H.A. Bernheim, B. Beutler, A. Cerami, I.S. Figari, M.A. Palladino, Jr., and J.V. O'Connor. 1986. J. Exp. Med., 163:1433-1450.
32. Elias, J.A., K. Gustilo, W. Baeder and B. Freudlich. 1987. J. Immunol. 138:3812-3816.
33. Renz, J., J.-H. Gong, A. Schmidt, M. Nainand D. Gemsa. 1988. J. Immunol., 141:2388-2393.
34. Chun, M. and M.K. Hoffman. 1987. Cancer Res. 47:115-118.

BACTERIAL ENDOTOXINS: RELATIONSHIPS BETWEEN CHEMICAL STRUCTURE

AND BIOLOGICAL ACTIVITY OF THE INNER CORE-LIPID A DOMAIN

Ernst Th. Rietschel, Lore Brade, Ulrich Schade, Ulrich
Seydel, Ulrich Zähringer, Otto Holst, Hella-Monika Kuhn,
Vladimir A. Kulschin, Anthony P. Moran, and Helmut Brade

Forschungsinstitut Borstel, Institut für Experimentelle
Biologie und Medizin, D-2061 Borstel, FRG

INTRODUCTION

Gram-negative bacteria such as the *Enterobacteriaceae* and
Pseudomonadaceae express at their surface various amphiphilic
macromolecules among which the endotoxins are of special microbiological, immunological and medical significance. Endotoxins are
essential for the organization and function of the bacterial
outer membrane, and, thus, for bacterial growth and survival. As
surface structures, endotoxins represent the main immunoreactive
antigens (O-antigens) of gram-negative bacteria, and they are
involved in the binding of antibodies and nonimmunoglobulin serum
factors, and, thus, in the specific recognition and elimination
of bacteria by the host organism's defense system. Further,
endotoxins are endowed with a broad spectrum of biological
(endotoxic) activities, such as pyrogenicity and lethal toxicity,
and they contribute to the pathogenic potential of gram-negative
bacteria. Finally, endotoxins activate B-lymphocytes and mononuclear cells and are potent immunostimulators. By virtue of
their biological activities they also seem to be involved in
certain physiological host-parasite interactions.

Endotoxins of various bacterial families possess a common
architecture (1). They consist of a heteropolysaccharide (O-
specific chain and core) and a lipid portion (termed lipid A)
and, hence, are chemically lipopolysaccharides (LPS). The polysaccharide region contains the structures involved in the binding
of O- and R-specific antibodies, lectins, bacteriophages, and an
ubiquitous complement-activating serum factor (1,2). In addition,
it exhibits a low degree of immunostimulatory activity (3). The
lipid A component, on the other hand, harbours the determinants
responsible for the induction of typical endotoxic effects such
as fever, Shwartzman reactivity and lethal toxicity (4). It
possesses strong immunostimulatory (Interleukin 1 (1_1-1)- and
tumor necrosis factor α (TNFα-inducing) activity (5,6,7) and
engenders and recognizes lipid A-specific antibodies (8,9). Both
lipid A and parts of the core are necessary for bacterial viability.

Microbial Surface Components and Toxins in Relation to Pathogenesis
Edited by E.Z. Ron and S. Rottem, Plenum Press, New York, 1991

209

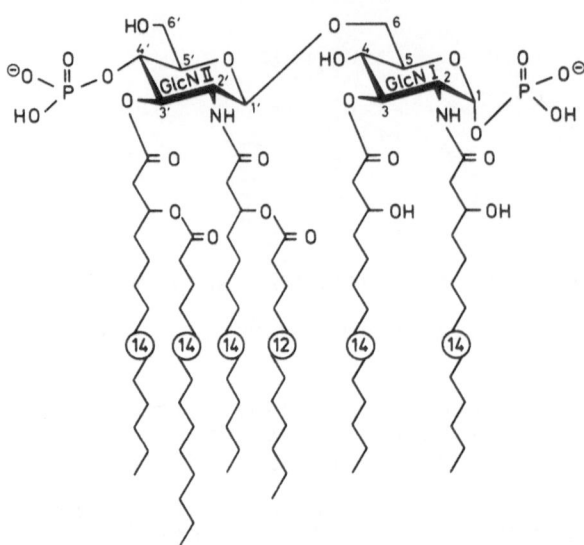

Fig. 1. Chemical structure of the lipid A component of *Esche-richia coli* (12)

    With the aim of establishing structure-activity relation-
ships on a molecular level we have investigated the chemical
structure as well as the biological and serological properties of
the lipid A component and the lipid A-proximal region of the core
oligosaccharide (termed the inner core).

LIPID A

    Enterobacterial lipid As have been studied most intensively
and, as an example, Fig. 1 shows the chemical structure of
Escherichia coli lipid A (10-12). It consists of a β(1-6)-linked
D-glucosamine disaccharide which carries two phosphoryl residues
in positions 1 and 4' (hydrophilic lipid A backbone), and four
(R)-3-hydroxymyristoyl groups in positions 2, 3, 2' and 3'. The
hydroxy fatty acids in positions 2' and 3' are, at their 3-
hydroxyl groups, acylated by lauric and myristic acid, respectively. (The α-linked phosphoryl group in position 1 may carry
nonstoichiometric amounts of phosphate). The hydroxyl group in
position 4 is free, and that in position 6' serves as the attachment site of the core oligosaccharide in intact LPS.

    In other bacterial families structural variants of *E. coli*
type lipid A have been encountered (for summary see (13)). These
variations concern e.g. the type and chain lengths of fatty acids
as well as their location at the backbone (13,14). Notably in
phototrophic bacteria, D-glucosamine may be replaced by 2,3-
diamino-2,3-dideoxy-D-glucose (15) and also differences in the
phosphorylation pattern of the backbone have been encountered
(16). Despite of these variations, the lipid A component is the
least variable region of biologically active LPS and its structure appears to be highly conserved. Further, lipid A represents
an obligatory constituent of LPS of different bacterial origin.

    Based on the results of chemical analysis, E.coli type lipid
A has been chemically synthesized (17) and shown to be physico-

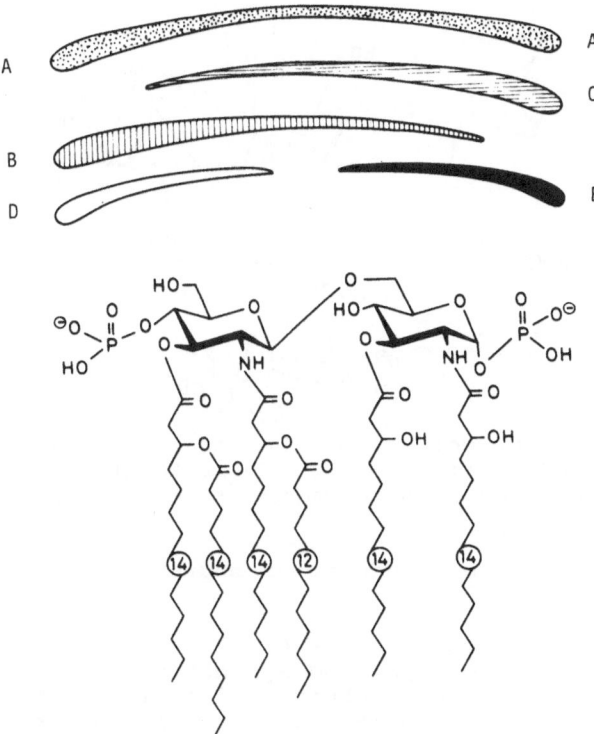

Fig. 2. Schematic display of the epitope specificities of disaccharide (A, B, C) and monosaccharide (D, E)-reactive anti-lipid A anti-bodies (20,21)

chemically and biologically identical to its bacterial counter-part (4,18). These findings proved that the previous structural proposal for lipid A is correct and that the endotoxic activity of LPS is principally mediated by or dependent on lipid A.

In attempts to define those minimal structural prerequisites which determine endotoxic activity of lipid A, we and others performed biological analyses using chemically modified bacterial lipid A, natural precursors of lipid A biosynthesis as well as chemically synthesized partial structures and analogues of lipid A. In particular, we have evaluated the pyrogenic, lethal, Il-1 and TNFα-inducing, and Shwartzman-phenomenon-provoking properties of synthetic preparations (4,5,6,7). Our results indicate that for the expression of the full spectrum of characteristic endo-toxic manifestations the simultaneous presence of the disaccha-ride, two phosphoryl groups, and optimally six acyl residues (including one 3-acyloxyacy group) in a defined distribution - as present in *E. coli* lipid A - are essential. Compounds lacking any of the constituents of this common structure of analogues with a different location of constituents, exhibit diminished or no endotoxicity. This shows that endotoxic activity is not deter-mined by one single lipid A constituent, i.e. a toxophore group, but rather depends on a defined overall chemical architecture

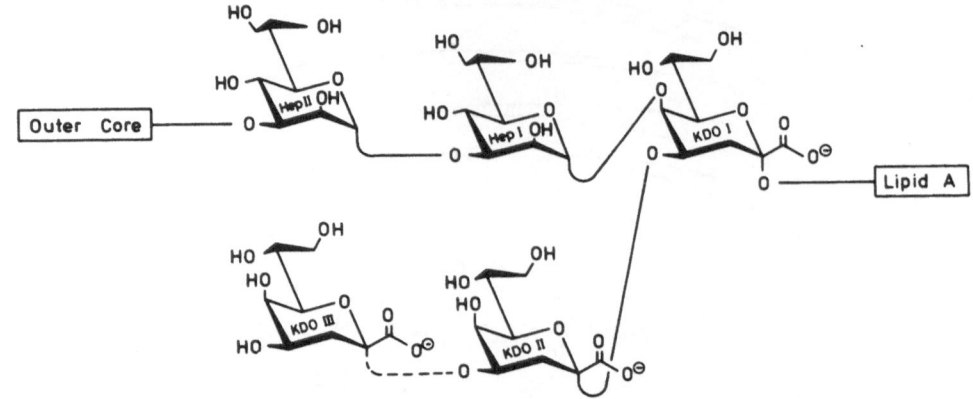

Fig 3.   Chemical structure of the inner core region of *Salmonella minnesota* LPS (9). Substituents such as phosphate
and phosphorylethanolamine are not shown.

which governs a unique molecular or supramolecular ("endotoxic")
conformation (13,18,19).

Immunoreactive determinants of *E. coli* type lipid A were
characterized with bacterial and synthetic lipid A antigens and
with the aid of polyclonal rabbit anti-lipid A antisera as well
as murine monoclonal antibodies. So far, five different antigenic
determinants have been defined, all residing in the hydrophilic
lipid A backbone region (20,21). One epitope comprises the 1,4'-
bisphosphorylated   β(1-6)-linked glucosamine disaccharide, a
second the 4'-phosphorylated, and a third the 1-phosphorylated
disaccharide (Fig. 2). The other two determinants are located in
acylated glucosamine 1- or 4-phosphate monosaccharides (Fig. 2)
Although in the expression of the different epitopes the presence
of phosphate groups is required, it cannot be decided at present
whether they are integral components of the determinants or
whether they generate or stabilize an immunoreactive conformational state of lipid A.

INNER CORE

The structure of the inner core region which is characterized by the presence of the unusual sugar acid 3-deoxy-D-manno-2-
octulosonic acid (KDO) has been elucidated in LPS of phosphate-
less *Salmonella minnesota* mutants, and it is shown in Fig. 3 (9).
The KDO-group (KDO I) which is bound to lipid A is substituted in
position 4 by a second KDO-residue (KDO II) or, in some strains
by a (2-4)-linked KDO-disaccharide. In position 5, KDO I carries
a (1-3)-linked disaccharide of L-glycero-D-manno-heptose. All
sugars of the inner core are present as α-pyranosides.

The structural arrangement of the inner core of other
enterobacterial and nonenterobacterial strains is less well
known, but it appears that at least one KDO residue represents an
obligatory constituent of LPS. In *Chlamydia trachomatis* (wild
type) LPS, the saccharide portion consists exclusively of KDO (in
the form of a trisaccharide). In this case, a lipid A proximal
α(2-4)-linked KDO disaccharide (KDO I and KDO II of Fig. 3)

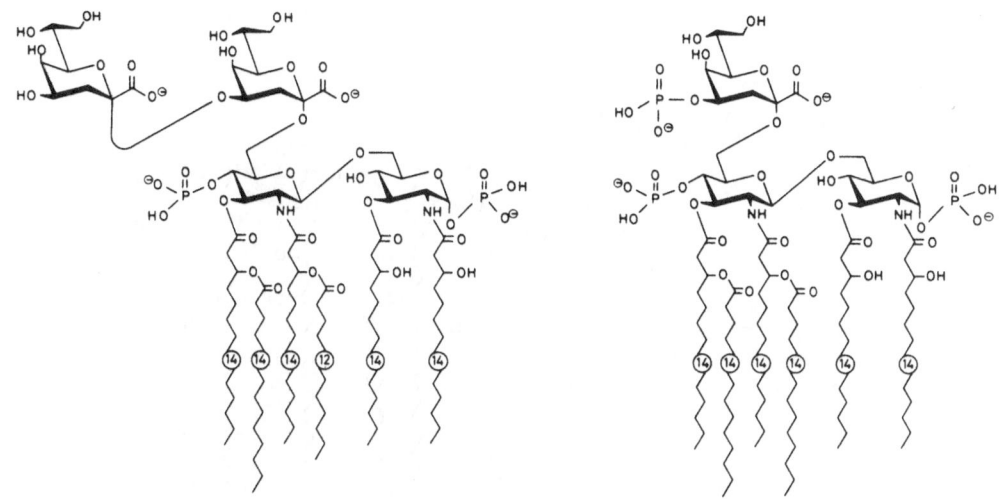

Fig. 4.   Minimal LPS structures of viable gram-negative bacterial (left: *Escherichia coli* Re mutant, strain 515 (12);
right: *Haemophilus influenzae* R-mutant, strain I-69
Rd-/b+ (24)).

carries at the primary hydroxyl group (C8) of KDO II a third α-
bound pyranosidic KDO-group (22,22a).

In other bacterial groups, KDO derivatives (such as KDO
phosphate and KDO pyrophosphoryl ethanolamine, for literature see
(23,24)) or KDO related sugars like D-glycero-D-talo octulosonic
acid (25), 3-deoxy-heptulosaric acid (26) and 3-deoxy-D-threo
hexulosonic acid (27) may be present in the inner core. Neuramin-
ic acid which is chemically related to KDO, is present in the LPS
of purple nonsulfur bacteria (28) and was recently also detected
in LPS of pathogenic *Campylobacter jejuni* bacteria (A. P. Moran
and U.Zahringer, unpublished results).

Immunoreactive determinants of the inner core have recently
been characterized (29) with monoclonal anti LPS antibodies (30)
using KDO-containing bacterial and synthetic antigens (31,32).
One antibody was found to be specific for a terminal α-linked KDO
residue (such as KDO III of Fig. 3). This antibody reacted with a
variety of enterobacterial LPS (29). A second antibody was
identified to be specific for an α(2-4) linked KDO disaccharide
as present in molar amounts in LPS of Re chemotypes of *S. minnesota*, *E. coli* and *P. mirabilis* and in strain R345 (Rb2 chemo-
type) of *S. minnesota*. Further studies concerned a nonimmunoglobulin 28 KDa factor which is present in sera of all mammals
including humans. This protein binds to a KDO containing determinant expressed by LPS of different bacterial families (2).

The combined serological results show that at least one KDO
residue of the inner LPS core is exposed and accessible to
recognition by humoral factors of the defense system.

KDO-LIPID A DOMAIN

Above, the structure and biological properties of the KDO con-

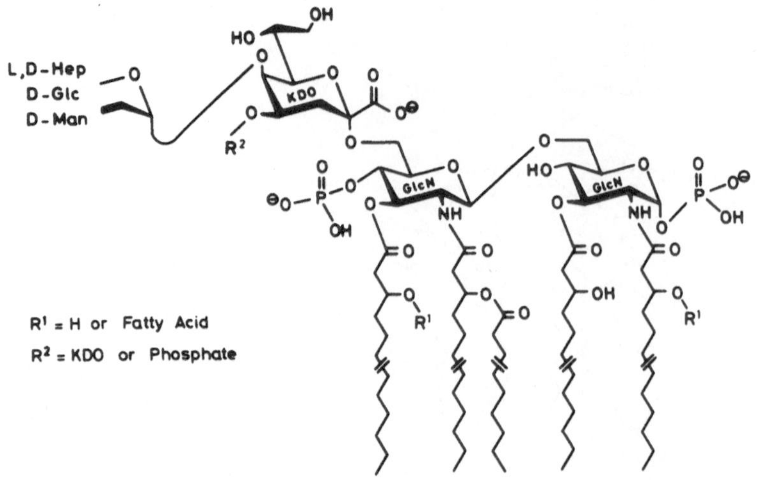

Fig 5.   Conservative region of biologically active LPS. The
           KDO- lipid A domain with attached substituents is
           shown.

taining inner core and the lipid A component have been described
separately. Recent evidence, however, suggests that in some
systems biological activity is optimally expressed by a molecule
containing both, KDO and lipid A. For example in the ability to
stimulate the release of leukotrien C4 from mouse peritoneal
macrophages (33) and of Il-1 or TNFα from human peripheral
monocytes (6,7), free lipid A was significantly less active that
LPS of enterobacterial Re-mutants (the latter consisting of lipid
A and a KDO disaccharide; KDO I and KDO II of Fig. 3).

     The KDO lipid A domain is also essential for bacterial
viability. So far, multiplying gram negative bacteria which lack
LPS have not been isolated. This points to a vital role of LPS
for bacterial growth and division. Since bacterial Re mutants,
the LPS of which consists of only lipid A and KDO (Fig. 4), do
multiply (at least *in vitro)* and since KDO defective mutants are
not viable, it appears that the O-specific chain, the outer core
and the heptose region are dispensable, while the innermost KDO
containing region of the core, together with lipid A, are essential for bacterial vitality. Until recently the minimal LPS
structure required for the structural and functional integrity of
the bacterial cell was believed to be that of (enterobacterial)
Re mutants (Fig. 4). In this context the *Haemophilus influenzae*
deep rough mutant (strain I-69 Rd-/b+) which was genetically constructed by Moxon et al. (34) is of significance. We have structurally characterized its LPS and shown that it contains only one
pyranosidic KDO residue which is phosphorylated and α- linked to
lipid A (24). Since this *H. influenzae* mutant is able to multiply
it follows that one (phosphorylated) KDO group in LPS, together
with lipid A, suffices for growth and survival of a gram-negative
bacterium.

     Fig. 5 shows the KDO-lipid A domain of LPS with attached
substituents. This domain represents the structurally most
conserved region of LPS. It is present in taxonomically remote
bacteria and it harbours ubiquitous immunoreactive epitopes. It
is this structure which is required for bacterial viability and

which expresses optimal activity in certain endotoxin test systems. It must be emphasized that the biological activity of the KDO-lipid A domain clearly depends on the presence of the lipid A component. The KDO-containing inner core *per se* is devoid of toxicity but it seems to contribute to lipid A mediated effects and, thus, to modulate lipid A activity. As our previous results show, the endotoxic, immunostimulatory and serological activities of lipid A depend on a unique supramolecular conformation which is determined by a peculiar primary lipid A structure and the fluidity of its acyl chains on the one hand, and the physicochemical environment on the other (13,18,19,21,35,36). We presently favour the concept that the KDO containing inner core influences the physical structure of lipid A resulting in a conformation which allows optimal display of lipid A activity. Using physical and serological techniques, we currently are aiming at the characterization of the conformation of the KDO lipid A domain. These studies are performed in the hope that knowledge of the supramolecular structure of this domain will enable us to understand, at the molecular level, the initial steps of the endotoxin - host interaction and, thus, the mechanism involved in endotoxin activity.

ACKNOWLEDGMENTS

The financial support of the Deutsche Forschungsgemeinschaft (Br 731/4-1, Br 731/7-1, Scha 402/1-1, Scha 402/1-2), the Bundesministerium fur Forschung und Technologie (HB, 01 ZR 8604), the Alexander von Humboldt Foundation (APM), and the Fonds der Chemischen Industrie (EThR) is greatfully acknowledged. We also thank Mrs. M. Lohs, G. Stegelmann, and B. Kohler for illustrations and photographic work, and Mrs. I. Bendt for typing this manuscript.

REFERENCES

1. Luderitz, O., M.A. Freudenberg, C. Galanos, V. Lehmann, E.Th. Rietschel, and D. Shaw. 1982. In: Microbial membrane lipids (S. Razin and S. Rottem, eds.). Academic Press Inc., New York 17:79-151.
2. Brade, L., and H. Brade. 1985. Infect. Immun. 50:687-694.
3. Lebbar, s., J.-M. Cavaillon, M. Caroff, A. Ledur, H. Brade, R. Sarfati, and N. Haeffner-Cavaillon. 1986. Eur. J. Immunol. 16:87-91.
4. Galanos, C., O. Luderitz, E.Th. Rietschel, O. Westphal, H. Brade, L. Brade, M.A. Freudenberg, U. Schade, M. Imoto, S. Yoshimura, S. Kusumoto, and T. Shiba. 1985. Eur. J. Biochem. 148:1-5.
5. Loppnow, H., L. Brade, H. Brade, E.Th. Rietschel, S. Kusumoto, T. Shiba and H.-D. Flad. 1986. Eur. J. Immunol. 16:1263-1267.
6. Loppnow, H., H. Brade, I. Durrbaum, C.A. Dinarello, S. Kusumoto, E.Th. Rietschel, and H.-D. Flad. 1989. J. Immunol. 142:3229-3238.
7. Feist, W., J. Musehold, H. Brade, A.J. Ulmer, S. Kusumoto, and H.-D. Flad. 1989. Immunobiol. 179:293-307.
8. Galanos, C.. M.A. Freudenberg, F. Jay, D. Nerkar, K. Veleva, H. Brade and W. Strittmatter. 1984. Rev. Infect. Dis. 6:546-552.
9. Brade, H., L. Brade and E. Th. Rietschel. 1988. Zbl. Bakt. Hyg. A. 268:151-179.

10. Seydel, U., B. Lindner, H.-W. Wollenweber and E. Th. Rietschel. 1984. Eur. J. Biochem. 145:505-509.
11. Imoto, M., S. Kusumoto, T. Shiba, H. Naoki, T. Iwashita, E.Th. Rietschel, H.-W. Wollenweber, C. Galanos and O. Luderitz. 1983. Tetrahedr. Lett. 24:4017-4020.
12. Zahringer, U., B. Lindner, U. Seydel, E.Th. Rietschel, H. Naoki, F.M. Unger, M. Imoto, S. Kusumoto and T. Shiba. 1985. Tetrahedr. Lett. 26:6321-6324.
13. Rietschel, E.Th., L. Brade, U. Schade, U. Seydel, U. Zahringer, S. Kusumoto and H. Brade. 1988. In: Surface structures of microorganisms and their interactions with the mammalian host. E. Schrinner, M.H. Richmond, G. Seibert and U. Schwarz, eds. pp. 1-40, Verlag Chemie, Weinheim.
14. Takayama, K., N. Qureshi, K. Hyver, J. Honovick, R.J. Clotter, P. Mascagni and H. Schneider. 1986. J. Biol. Chem. 261:10624-10631.
15. Mayer, H. and J. Weckesser. 1988. FEMS Microbiol. Rev. 54:143-154.
16. Weintraub, A., U. Zahringer, H.-W. Wollenweber, U. Seydel and E.Th. Rietschel. 1989. Eur.J.Biochem. 183:425-431.
17. Imoto, M., H. Yoshimura, T. Shimamoto, N. Sakaguchi, S. Kusumoto and T. Shiba. 1987. Bull. Chem. Soc. Jpn. 60:2205-2214.
18. Rietschel, E.Th., H. Brade, K. Brandenburg, U. Schade, U. Seydel, U. Zahringer, C. Galanos, O. Luderitz, O. Westphal, H. Labischinski, S. Kusumoto and T. Shiba. 1987. In: Detection of bacterial endotoxins with the LAL test. S.W. Watson, J. Levin, Th.J. Novitsky, eds. Progr. Clin. Biol. Res. 231:25-53. Alan R. Liss, Inc. New York.
19. Labischinski, H., D. Naumann, C. Schultz, S. Kusumoto, T. Shiba, E.Th. Rietschel and P. Giesbrecht. 1989. Eur. J. Biochem. 179:659-665.
20. Brade, L., E.Th. Rietschel, S. Kusumoto, T. Shiba and H. Brade. 1986. Infect. Immun. 51:110-114.
21. Brade, L., K. Brandenburg, H.-M. Kuhn, S. Kusumoto, I. Macher, E.Th. Rietschel and H. Brade. 1987. Infect. Immun. 55:2636-2644.
22. Brade, H., L. Brade and F.E. Nano. 1987. Proc. Natl. Acad. Sci. USA 84:2508-2512.
22a. Kosma, P., G. Schulz and H. Brade. 1988. Carbohydr. Res. 183:183-199.
23. Brade, H. 1985. J. Bacteriol. 161:795-798.
24. Helander, I., B. Lindner, H. Brade, K. Altmann, A.A. Lindberg, E.Th. Rietschel and U. Zahringer. 1988. Eur. J. Biochem. 177:483-492.
25. Kawahara, K., H. Brade, E.Th. Rietschel and U. Zahringer. 1987. Eur. J. Biochem. 489:489-495.
26. Brade, H. and E.Th. Rietschel. 1985. Eur. J. Biochem. 153:249-254.
27. Kondo, S., U. Zahringer, E.Th. Rietschel and K. Hisatsune. 1989. Carbohydr. Res. 188:97-104
28. Krauss, J.H., G. Reuter, R. Schauer, J. Weckesser and H. Mayer. 1988. Arch. Microbiol. 150:584-589.
29. Brade, L., P. Kosma, B.J. Appelmelk, H. Paulsen and H. Brade. 1987. Infect. Immun. 55:462-466.
30. Appelmelk, B.J., V. van Vught, H. Brade, J. Maaskant, W. Schouten, B. Thijs and D. MacLaren. 1988. Progr. Clin. Biol. Res. 272:373-382 (Alan R. Liss, Inc. New York.
31. Paulsen, H. and M. Schuller. 1987. Liebigs Ann. Chem. 167:249-258.

32. Kosma, P., J. Gass, G. Schulz, R. Christian and F.M. Unger. 1987. 167:39-54.
33. Luderitz, O., K. Brandenburg, U. Seydel, A. Roth, C. Galanos and E.Th. Rietschel. 1989. Eur. J. Biochem. 179:11-16.
34. Moxon, E.R. 1985. In:Bayer-Symposium VIII. The pathogenesis of bacterial infections, G.G. Jackson and H. Thomas, eds. pp.17-29, Springer Verlag, Berlin, Heidelberg.
35. Brandenburg, K. and U. Seydel. 1984. Biochem. Biophys. Acta. 775:225-238.
36. Brandenburg, K. and U. Seydel. 1985. Thermochim. Acta. 85:473-476.

SYMPOSIUM HELD UNDER THE AUSPICES OF THE FEDERATION OF EUROPEAN MICROBIOLOGICAL SOCIETIES
MAY 15 - 19, 1989
RAMAT RACHEL, ISRAEL

AUTHOR INDEX